THE COMMONWEALTH AND INTERNATIONAL LIBRARY
Joint Chairmen of the Honorary Editorial Advisory Board
SIR ROBERT ROBINSON, O.M., F.R.S., LONDON
DEAN ATHELSTAN SPILHAUS, MINNESOTA
Publisher: ROBERT MAXWELL, M.C., M.P.

LIBERAL STUDIES DIVISION
General Editors: E. F. CANDLIN AND D. F. BRATCHELL
SCIENCE AND SOCIETY VOLUME 6

LATE SEVENTEENTH CENTURY SCIENTISTS

LATE SEVENTEENTH CENTURY SCIENTISTS

EDITOR:
DONALD HUTCHINGS
DEPARTMENT OF EDUCATION
UNIVERSITY OF OXFORD

PERGAMON PRESS
OXFORD · LONDON · EDINBURGH · NEW YORK
TORONTO · SYDNEY · PARIS · BRAUNSCHWEIG

Pergamon Press Ltd., Headington Hill Hall, Oxford
4 & 5 Fitzroy Square, London W.1

Pergamon Press (Scotland) Ltd., 2 & 3 Teviot Place, Edinburgh 1

Pergamon Press Inc., Maxwell House, Fairview Park, Elmsford, New York 10523

Pergamon of Canada Ltd., 207 Queen's Quay West, Toronto 1

Pergamon Press (Aust.) Pty. Ltd., 19a, Boundary Street, Rushcutters Bay, N.S.W. 2011, Australia

Pergamon Press S.A.R.L., 24 rue des Écoles, Paris 5e

Vieweg & Sohn GmbH, Burgplatz 1, Braunschweig

First edition 1969
Library of Congress Catalog Card No. 68-57233

Printed in Great Britain by Hazell Watson & Viney Ltd., Aylesbury, Bucks

08 013358 4 (flexicover)
08 013359 2 (hard cover)

CONTENTS

ACKNOWLEDGEMENTS

THE publisher and the authors acknowledge with thanks the following publishers and authorities who have granted permission to quote from publications of which they hold the copyright: Hollandsche Maatschappij der Wetenschappen (*Oeuvres Complètes de Christiaan Huygens*, Société Hollandaise); Cambridge University Press and Harvard University Press (Isaac Newton's Papers and letters on Natural Philosophy); Executors of the late Dr. R. T. Gunther and Oxford University Press (Gunther: *Early Science in Oxford*).

INTRODUCTION

THE most important aspect of the study of science lies in the understanding of science itself. It was mainly in the sixteenth and seventeenth centuries that men came to recognize science as an intellectual discipline thoroughly based in experience.

The Renaissance, which by the sixteenth century was well under way in Italy, soon spread to northern Europe, first to Germany, then France and the Low Countries, and eventually to England. At first, the revival was mainly literary, but gradually men began to pay less attention to what was written in ancient books, like those of Aristotle, and instead placed more reliance on their own observations. Natural philosophers—scientists we should now call them—were less inclined to speculate about metaphysical problems but concentrated their attention on questions that could be investigated under simplified controlled conditions. A hypothesis led to a new experiment and this in turn to a new hypothesis.

By the middle of the seventeenth century considerable progress had been made in mechanics, astronomy, and optics. The importance of all this remarkable scientific work lies not so much in the discoveries themselves but in the intellectual processes and experimental methods which brought them into being. That comprehension of natural phenomena should conform with experimental findings and rigorous, logical reasoning. New mathematical techniques became accepted as a fundamental tool of science, and several important scientific instruments, including the telescope, the microscope, the air-pump, and the pendulum clock, were perfected, enabling more precise observation and measurement.

It was to foster the experimental philosophy that the Royal Society came into being in 1660. Soon given support by Charles II, it received its Royal Charter in 1662. The early members were scientifically minded men, themselves usually amateurs, who put across their ideas to each other and to those members like Samuel Pepys and the king himself whose presence was motivated more by curiosity than a consuming interest in science. The Society does not seem to have suffered because of its "non-playing" members, and the meetings gave plenty of opportunity for exchanging ideas and criticizing each others work.

All six essays in this volume are concerned with the lives and scientific work of men who were active in the latter half of the seventeenth century. They include some of the greatest names in the history of man's intellectual achievement. All six were members or were otherwise connected with the Royal Society.

The first of our subjects, Robert Boyle, was one of the earliest members. He demonstrated his air-pump, and his experiment attempting to weigh air seemed ridiculous to the non-scientists present—Charles II joined in the laughter. But, as the first essay describes, it is mainly as a chemist that Boyle should be remembered. He pioneered physical methods, but his greatest contribution to scientific understanding was his insistence that a substance should be regarded as an element until it can be further resolved into simpler substances. The objects of the Society were drawn up by Robert Hooke. He was recommended to the Society by Boyle as Curator and Experimenter. His duty was to furnish experiments to be shown at the weekly meetings. He laid down the Society's attitude to excessive or dogmatic theorizing :

"[it] will not own any hypothesis, system or doctrine of the principles of naturall philosophy, proposed or mentioned by any philosopher ancient or modern, nor the explication of any phenomena whose recourse must be had to originall causes (as not being explicable by heat, cold, weight, figure, and the like, as effects produced; thereby); nor dogmatically define, nor fix axioms of scientificall things, but will question and canvas all opinions, adopting nor adhering to none, till by mature debate and clear arguments chiefly such as are deduced from legitimate experiments, the truth of such experiments to be demonstrated invincibly."

Yet, as the essays on Hooke and Newton show, it was Hooke who found fault with Newton's experimentally based theory of light and colours, arguing against it and in favour of his own "wave theory".

Another of the objects of the Royal Society was "To improve the knowledge of . . . all useful Arts, Manufactures, Mechanick practices, Engynes and Inventions by Experiments. . . ."

In other words, the Society was as much concerned with technology as what we should now call *pure* science. The essay on Christopher Wren demonstrates the versatility of most seventeenth-century scientists. Wren was exceptional, we are told, not because he took up architecture in addition to mathematics and astronomy, but because of his great success as an architect!

Our fourth subject is Christian Huygens, the Dutch mathematician and scientist, who first solved the problem of "centrifugal force" required to keep a body moving in a circle with constant speed. Among his many contributions to scientific understanding, his theory of light is still taught in its original form.

The outstanding achievements of physical science in the seventeenth century tend to overshadow all else. But, as Dr. Wilkie reminds us, the attitude of science to living things was also changing rapidly. In the life and work of Italian Marcello Malpighi we have an example of biological method and reasoning, including the employment of models as explanatory devices.

The climax of the intellectual revolution was Isaac Newton's synthesis of all problems of motion, of the earth and of the sky, in his laws of motion and gravitation : this synthesis set a pattern for the mechanical view of the universe that was to dominate scientific thought for more than 200 years.

CHAPTER 1

ROBERT BOYLE, 1627–1691

D. C. Firth

Robert Boyle was born in Lismore, Ireland, in January 1627. He was the seventh son and fourteenth child of Sir Robert Boyle who was the Earl of Cork and Lord High Treasurer of Ireland. The financial security afforded by estates in Ireland and in England allowed Boyle the freedom to pursue his scientific work, gave him access to court circles which proved so valuable to the early Royal Society, but at the same time imposed limitations to which he so often refers. Public work involved meetings and travel, there were regular visitors to his laboratories, and his vacuum pump seems to have been one of the sights of London. Demonstrations of his "pneumatical experiments" were regarded by many as fine entertainment, and a useful comparison might be drawn between them and some of the work carried out at the Royal Institution over a hundred years later. It is hard to assess the value of public interest of this kind, but we might suppose it to be of lasting value to science. Boyle's eldest sister, Katherine, proved to be a most important influence on him, his respect for her was great, and through her he came into contact with some of the men who were responsible for founding the English scientific societies.

His early education began at Eton. He was a pupil there for 4 years, starting at the age of 8. Afterwards he was sent to the Continent and he lived abroad for 6 years, during which time he travelled widely and gained a lasting interest in foreign languages. Whilst abroad his studies included French and Italian, mathematics and geography, and his physical education included fencing and dancing. His reading included some of Galileo's work and a little modern history. Many children with his background were educated in this way, and the value of a period spent abroad may be seen in their later work. In Boyle's case there remained a life-long interest in geography and travel, a heightened appreciation of his own language, and a familiarity and interest in the work of scientists from other countries.

He returned to England in 1644 under the cloud of the Irish Rebellion, which was threatening the family resources, and on his return he saw at first hand the parliamentary unrest which was dividing his family. His sister Katherine was living in London at this time, and a short stay with her brought him into contact with men who subscribed to the Parliamentary Cause. Later in the same year he moved to a family estate at Stalbridge in Dorsetshire, where he seems to have divided his time between farming, experimenting, writing philosophy, and corresponding with men of "the Invisible College". Boyle himself was responsible for the title, and it refers to an informal association, dedicated amongst other things to the ideals of the Czech educator, Comenius. The group will be described more fully when the Royal Society is discussed.

After spending some 6 years at Stalbridge, he travelled to Ireland. Reduced tension made possible the work of repairing his estates there, and Boyle's correspondence at the time shows his annoyance at the way administration of his business was interfering with the scientific activities. One interesting aspect of his visit to Ireland is that it was here he came into close contact with members of the Gresham group—the forerunner of the Royal Society. In 1654 Boyle moved to Oxford, and it was here that much of his best scientific work was done. It was here that he influenced the first members of the Royal Society, cultivating chemistry as a study which must take its place alongside physics and mathematics. After some 14 years in Oxford he moved to London, and he remained there until his death in 1691.

It was probably a Victorian who styled him the "Father of Chemistry", but much earlier than this chemists recognized the contribution he had made. Francisco Redi, an Italian philosopher of the early eighteenth century said of him : "he was the greatest man who ever was, and perhaps ever will be, for the discovery of natural causes". Epithets like these are difficult to justify, but if they mean anything at all, they carry with them the conviction that Boyle was not only a successful experimenter, but that he also created the image of chemistry which exists today. There is no disputing the fact that the quality and quantity of his experimental work was quite phenomenal, and, as will be seen later, it represents the work of an inspired research team. The work is also distinguished for its span, for the wide variety of phenomena studied. That he was responsible for our image of chemistry is not so easy to demonstrate. He had many of those attributes we admire in a chemist, but whether he was the original holder of them is doubtful. He used gravimetric techniques, but so did Van Helmont before him. It is said that he destroyed the old concept of "The Elements" (see later), but for some the idea lived on, whilst for others it was apparently dead already.

It might not be too much of an exaggeration to claim that, in a century when the best scientific minds were preoccupied with mechanics and astronomy, and at a time when chemical knowledge rested uneasily between medical men, manufacturers, and alchemists, he made chemistry the concern of the major scientific societies both in England and on the Continent. This he did by applying the "Corpuscular Hypothesis" to explain chemical changes, by using physical techniques for the solution of chemical problems, and by publicly dissociating himself from the speculation, arguments, books of doubtful authenticity, and the aims of chemists of his period. Again, the accident of his birth must not be overlooked. All levels of society, including the Court, were accessible to him; he was assured of a respectful audience in this country at least.

THE REVIVAL OF THE "CORPUSCULAR HYPOTHESIS" AND THE BACONIAN IDEAL

During the seventeenth century philosophers reintroduced the view that matter was corpuscular in nature, the atom became a scientific device. Classical atomism taught a way of life, it indicated an attitude to all problems, it was an atheistic philosophy which could not possibly survive in a Europe dominated by Christian–Aristotelian thinkers. Now in the seventeenth century the atom was reborn to serve the limited purpose of explaining natural phenomena. Gassendi, a leading advocate of the theory, was a Catholic priest. Boyle and Newton both subscribed to the view and were devout Christians at the same time. The atom and the vacuum which surrounded it were both in harmony with the religious thinking of the times. The Roman poet Lucretius, *preached* atomism; Robert Boyle merely used it. The Roman says :

"... our starting point will be this principle : Nothing can ever be created by divine power out of nothing. The reason why all mortals are so gripped by fear is that they see all sorts of things happening on the earth and in the sky with no discernable cause, and these they attribute to the will of God.

"The second great principle is this : nature resolves everything into its component atoms and never reduces anything to nothing." (Penguin Classic, Lucretius, *The Nature of the Universe.*)

Sixteen hundred years later Newton wrote at the end of the *Opticks*; "It seems probable to me that God in the Beginning formed matter in solid, massy, hard, impenetrable, movable particles, of such size and figures, and with such properties and such proportions to space, as most conduced to the End for which he formed them. . . ."

It would be wrong to say that Lucretius preached atheism and that the atom was incidental. He does, in fact, describe their shapes and behaviour in great detail, and little was added to his description by the seventeenth-century scientists. As scientific hypotheses, little separates the two views, except perhaps a change of emphasis. The classical atom behaved by virtue of its shape, the new atom behaved because it had mass and speed. Great advances in dynamics, and the successful application of this to astronomy, created a climate in which the behaviour of atoms must be explained in the same terms.

During the century two major corpuscular theories emerged, one promoted by Gassendi, the other by Descartes, and there was some dispute between the protagonists of the different systems. It is significant that Boyle made little contribution to the structure of either theory, and that he was active in trying to minimize their difference. His approach to this problem reflects his general attitude to theory and its place in science; on the one hand, lies his conviction that ill-informed speculation is destructive to the cause of science, carrying with it a disregard for the facts, whilst, on the other hand, we have his own natural desire to generalize.

Gassendi regarded the atom as an ultimate particle, and in this he may be regarded as a descendant of the classical atomists. For him the atom was indestructible, and it moved about in a vacuum, the latter being a logical extension of the former. If an atom moves then there must be space elsewhere. In the beginning God gave the atom motion and the laws of mechanics did the rest. For Descartes there could be no vacuum, space was entirely filled. There were three kinds of matter, all corpuscular in nature, but they differed in the size of their particles. Terrestrial matter was composed of large atoms, spaces between these contained "ether" (smaller particles), and smaller particles still filled the remaining space. According to Descartes atoms were not absolute; they could be broken or worn down, and this process did occur. Boyle commented on the two views :

"I esteemed that, notwithstanding these things wherein the Atomists (followers of Gassendi) and the Cartesians (followers of Descartes) differed, they might be thought to agree in the main, and their hypotheses might, by a person of reconciling disposition, be looked upon as the matter, one philosophy. Which because it explicates things by corpuscles, or minute bodies, may not very unfitly be called corpuscular." (*Boyle's Wk.*, 1774, Vol. 1; brackets my own.)

In another place Boyle describes an experiment to investigate the Cartesian Ether, and before he describes his method he writes:

"I will not now discuss the Controversie betwixt some of the Modern Atomists, and the Cartesians. . . ."

Instead of argument he declares himself interested in attempting to detect the tenuous substance. The air-pump is used to free a space of air, the contents of the space are then expelled and the flow of ether is made to impinge on a feather. "If when our vessels are freed from air, they are full of Aether, that Aether is such a body, as will not be made sensibly to move a light feather. . . ."

In all his work Boyle attempted to realize the Baconian Ideal. According to this, facts must be collected and speculation must be repressed. The function of science was to be useful, it was to "improve man's estate". Boyle, in refusing to take sides in disputes of this kind, tried to give the impression that for him theory was of secondary importance; he tried so hard to be a Baconian. Reading between the lines there is no doubt that he found the position a great strain. He was probably very interested in current speculation, and may even have been pleased to see the feather remain unmoved.

He tries to follow Bacon when he writes about vacua, but here again it is a forced position; sitting on the fence is not easy for him:

". . . it was not my purpose to deliver my own opinions whether there be a Vacuum, or no, and though I do not in this Tract intend to declare myself either way; yet, that I may on this occasion also show, that the pressure of the air may suffice to account for divers phaenomena, which according to the vulgar Philosophers must be referr'd to Natures abhorrency of a Vacuum. . . ." (a continuation of *New Exp.*, 1669, p. 19.)

Boyle's desire to remain above a controversy on speculative matters caused him to remark that he learned his science at the hands of illiterate craftsmen, men who had no knowledge or interest in theory. He also said that he had for a time avoided reading the views of Gassendi and Descartes so that he could better judge things for himself.

We might consider this an artificial pose. On occasions he

demonstrates the breadth of his reading and his appreciation of current speculation, and when he does write in this vein his style is that of a very interested man. His constant plea, that unfounded theory should be confronted with experimental fact, sets him as a man apart. His occasional indulgence in verbal slanging sets him a little below the angels.

"Methinks the Chymists, in their searches after truth, are not unlike the navigators of Solomons Tarshish Fleet, who brought home from their long and perilous voyages, not only gold, and silver, and ivory, but apes and peacocks too : for so the writings of several (I say not all) of your hermetic philosophers presents us, together with diverse substantial and noble experiments, theories, which either like peacock's feathers make great show, but are neither solid nor useful, or else like apes, if they have some appearance of being rational, are blemished with some absurdity or other. . . ." (*Skeptical Chymist*, Everyman's Library, p. 227.)

Boyle's criticism of current theory was not always of this kind. Often it was rich in experiment and observation, but it is probably true to say that his references to speculation were usually destructive in character, and that his direct positive contribution to theory was very small. His failure (or success) to state his own opinions clearly, and the lack of conclusion in his own researches, did not inhibit the speculation of others. The Baconian Ideal was unworkable. Science progresses by initial speculation and then by experiment, the latter being inspired by the former. Like most men who persist in sitting on the fence, Boyle incurred the criticism of his contemporaries. Arrogance, humility, and ignorance are easily confused, they may all result in silence, and his silence was interpreted in these various ways. Huygens suggested that he was incapable of making generalizations.

The lack of theme and absence of general conclusion which characterized Boyle's work was characteristic of English science in general. Hooke's *Micrographia* is a bewildering collection of experiments and observations, each section being unrelated to those before and after. Bacon's influence was very strong in England, and the work of English scientists is in sharp contrast to that of the French, who were influenced by Descartes. French work of the period is characterized by a grand theme and by deductive reasoning, English work by an emphasis on experiment and the possible usefulness of discovery.

Boyle may be regarded as an unhappy Baconian. His attempts to practise the Ideal were a strain, and on occasions they ended in failure. His interest in current theory emerged at all points in his work, and the veneer of indifference to it is often transparent. Still, Boerhaave had no doubts about the origin of Boyle's inspiration: "Mr. Boyle, the ornament of his age and country, succeeded to the genius and inquiries of the great Chancellor Verulam (Bacon). Which of Mr. Boyle's writings shall I recommend? All of them. To him we owe the secret of fire, air, water, animals, vegetables, fossils: so that from his works may be deduced the whole system of natural knowledge."

Had Boyle tried to deduce "the whole system of natural knowledge" from his wide experience of materials, then his efforts would have proved a failure. No doubt Boyle knew this, and set himself a more limited objective.

THE CHEMICAL ELEMENTS AND THE BACONIAN IDEAL

The primary classification of substances into elements, compounds, and mixtures had a very slow evolution, and it could be argued that modern chemistry started when these ideas were formulated. The fact that they appear at the beginning of elementary texts reflects a method of teaching the subject; at the same time it hides their difficult origins. The claim that Boyle was the first modern chemist has been made on the grounds that he was responsible for these central concepts, and this aspect of his work must be considered.

When Boyle discusses the chemical theory of the day, he refers to two classes of opinion, that of the "hermetic philosophers" or "peripatetics" and that of the "spagyrists", "vulgar spagyrists", or "chymists". The hermetic philosophers, or followers of Aristotle, held that there was one matter from which all things were made, and that four properties or qualities were imposed on it in various degrees. The four qualities were hotness, coldness, wetness, and dryness; and one material only differed from another in the degree to which it contained these. According to this philosophy there were four elements, and these were the limiting cases:

Element	*Qualities*
FIRE	HOTNESS + DRYNESS
AIR	HOTNESS + WETNESS
WATER	COLDNESS + WETNESS
EARTH	COLDNESS + DRYNESS

The many changes to be seen in nature occurred because of each of these elements had a natural resting place. Fire rose because it tended to that part of the universe which was right for it, earth fell to the ground because its natural place was the centre of the universe (the Earth).

The peripatetic and the experimental philosopher (Boyle) could have little contact. Aristotle's biological studies show that he was a tireless observer, and that he had an intense curiosity towards the things about him. Unlike their master, the peripatetics had little need for observation. Their acceptance of a philosophy rested, not on what they saw, but upon a respect for scholars of the past. For them, individual experiments were trivial and irrelevant; truth was to be found in history, and the more ancient the authority the better it was. In *The Skeptical Chymist*, Boyle recognizes the futility of presenting experiments to the thoroughgoing peripatetic, and he chooses an easier opponent, a man called Themistius, who is prepared to justify the peripatetic view of matter by referring to observations! Poor Themistius begins his defence of the "Four Elements" by complaining:

"I engage in this controversie with great and peculiar disadvantages. . . . For he [Boyle] justly apprehending the force of truth, has made it the chief condition of the duel, that I should lay aside the best weapons I have, and those I can best handle (knowledge of and respect for authority.)

". . . and I hope you will however consider, that that great favourite and interpreter of nature, Aristotle, who was the greatest master who ever lived, disclaimed the course taken by other petty philosophers, who not attending the coherence and consequences of their opinions, are more solicitous to make each particular opinion plausible independently upon the rest." (*The Skeptical Chymist*, Everyman edn., p. 19.)

Peripatetics, used the term "element" in its Aristotelian sense; for them, water was an element, not because its matter was simple, but because its qualities or properties were elementary.

The "spagyrists" or "vulgar chemists" owed their origin to the teaching of Paracelsus, born about 1493 and a native of Zürich. His eventful career began as a student of medicine at Basle; he travelled widely in Europe, lectured at the University of Basle, was forced to leave there after crossing swords with the authorities, and finally settled in Salzburg under the protection of the Archbishop. He was one of the most outspoken sons of the Renaissance; classical medicine was his hatred. His vision of chemistry was that it must serve medicine, it must provide potions. More than this, he advocated a new philosophy of materials, proposing that there were three elementary qualities—mercurial, sulphurous, and salt like. These Paracelsan qualities are very similar to Aristotles' coldness, hotness, wetness, and dryness, and their imposition on matter, to form "mercury", "sulphur", and "salt", adds very little to chemical theory. "Sulphur" is equivalent to "fire", "salt" to "earth", and "mercury" to "water" or "air". The relationship of the Paracelsan elements to the three common chemicals bearing the same names is obscure, and a constant source of confusion. Some sixteenth-century chemists used the words in a purely adjectival sense; for them a substance may be sulphurous (inflammable); mercurial (volatile); salt-like (resistant to fire); or any combination of these. Other chemists regarded "mercury", "salt", and "sulphur" as ideals, and saw the common chemicals as approximations to these. For some chemists, the statement "paper is rich in sulphur", simply meant that paper is inflammable; for others it meant that paper actually contained an element called sulphur. Paracelsus himself was obscure about the meaning of the terms, and as time progressed the confusion deepened.

The Paracelsan nomenclature, and his ambition to make chemical studies a branch of medicine, were the greatest influences of the sixteenth century. Interest in new chemicals sprang from the possible medicinal use they might have. The Paracelsan elements became absorbed into the physiology of the times. Some disorders were due to an excess of "sulphur", some to a deficiency. This kind of chemistry has been called iatrochemistry, and among its many students may be listed Angelus Sala, Daniel Sennert, J. B. Van Helmont, Otto Tachenius, and J. R. Glauber. The latter is remembered for his preparation of Glauber's salt; the fact that he advanced it as a cure for almost every ailment is not so well known.

During this period fraudulent writings were common. A lesser known writer might publish his work under the name of one who had already won fame, hoping that it would reach a wider public. Styles were imitated but words were used in a different sense. Some took it upon themselves to translate the essential terms so that their meaning might be clearer. Glauber wrote :

"The principles of vegetables are water, salt, and sulphur, from which also the metals are derived, not from the running mercury, as many of you think, for that mercury is a special metal and from these same three principles as other metals and vegetables. . . ." (*De Natura salium opera chymica*, 452.)

Individual chemists used their own combinations of elements. Van Helmot reverted to a pre-Socratic idea that there was only one element—"water". Others worked on the basis of five or possibly six. One theory held that there were three active principles—spirit, oil, and salt, and two passive—water and earth.

"These are called active because being in active motion they cause the activity of the compound. The others are called passive because being in repose they serve only to restrain the vivacity of the active ones" (tr. from 9th Paris edn. of *Cours de Chymie*, 1701, N. Lémery).

Boyle's contribution to the problem of the elements lies in his insistence that there must be a common vocabulary, and in his thorough examination of the substances claimed to be elementary. Most important of all was his view that chemical changes could not be explained in terms of elements, and that they had no existence. His views on the subject are in *The Skeptical Chymist*, printed in English in 1661. Shortly afterwards Latin translations appeared, and there was an immediate and wide circulation. The book is written as a dialogue between Carneades (Boyle) and Themistius, a peripatetic. The spagyric view is represented, but Carneades usually addresses himself to the peripatetic, his remarks applying equally to all philosophers who insist on a limited number of elements. Early in the book Boyle attacks the current terminology. Carneades says :

"I have long observed that those dialectical subtleties, that the schoolmen too often employ about physiological mysteries, are wont much more to declare the wit of him that uses them, than increases the knowledge or remove the doubts of sober lovers of truth. And such captious subtleties do indeed often puzzle and sometimes silence men, but rarely satisfy them."

Briefly, Boyle's criticism of current vocabulary is this; it is not informative, it is designed to confuse people, it tells more of the writer than of his subject.

The peripatetic is persuaded to demonstrate the Four Element view, and he does this by describing the burning of a piece of wood. Fire is observed and this rises, smoke (air) is observed and this also rises, water appears at the cut surfaces, and ash (earth remains. Carneades then attacks the argument, and he does so by questioning the interpretations, not the observations. We may list the points he makes.

1. Why should fire be regarded as a method of analysis? It produces different results according to its intensity, according to whether the vessels used are open or closed. Sometimes it produces no resolution at all, as in the case of heating gold.
2. If fire does in fact resolve matter into its elements, how can it be argued that there are four? Some materials give five products on heating, and some less than four.
3. The products evolved when a substance burns might not exist in the substance beforehand, but might be created by the fire.

"I consider then in the first place there are divers bodies out of which Themistius will not prove in haste that there can be so many Elements as four extracted by the Fire. And I should perchance trouble him if I should ask what Peripatetic can show us (I say not all four Elements, for that would be too rigid a question) but any one of them extracted out of gold by any degree of Fire whatsoever,

". . . so there are others which may be reduced into more, as the blood (and divers other parts) of men and other animals, which yield when analysed five distinct substances. Phlegm, Spirit, Oyle, Salt, and Earth." (*The Skeptical Chymist*, p. 27.)

"And let me add, Philoponus, that to make your reasoning cogent, it must first be proved that the fire does only take the elementary ingredients asunder, without otherwise altering them. Fore else 'tis obvious, that bodies may afford substances which were not pre existent in them" (*The Skeptical Chymist*, Everyman edn., p. 24).

Boyle's own views about the elements are easily misunderstood. Throughout the book he argues about proof, but it does seem that he has sympathy with the central idea—that substances are composed of a limited number of elements and that it would be very useful to know what these are. The reader might be disturbed when, quite early in the discussion, Carneades presents a corpuscular view of matter, and he is little the wiser when the conversation then continues as before. It is not clear at the time that Boyle is presenting the corpuscular hypothesis as an *alternative* way of explaining chemical changes, and that he doubts the premise that there are elements. It is only in the sixth part of the book that this emerges.

"And, to prevent mistakes, I must now advertize you, that I now mean by elements, as those chymists that spent plainest do by their principles, certain primitive and simple, or perfectly unmingled bodies; which not being made of any other body, or of one another, are the ingredients of which all those called perfectly mixt bodies are immediately compounded, and into which they are ultimately resolved : now whether there be any one such body to be constantly met with in all, and each, of those text are said to be elemented, is the thing I now question."

The first sentence here, presents a modern appearance, and at a first reading, we might wonder why Boyle failed to anticipate the work of Lavoisier who provided us with a list of elements. A more careful reading shows why he failed to provide a list; his definition is quite different from the one Lavoisier used. Boyle took an element to be a substance which is present in *all* compound bodies, and on these grounds, sulphur, carbon, copper, and the rest, could not possibly be regarded as elements because they were not universal ingredients.

Boyle's contribution to the evolution of the modern element lies, not in being the first to use our idea, but in his attempt to destroy the old concept. He doubted the existence of universal ingredients, and chemists who accepted his point of view were then in a position to create the new notion.

"Chymists pretend from some (substances) to draw salt, from others running mercury, and from others sulphur; yet they have not taught us by any way in use among them to separate any one principle, whether salt, sulphur, or mercury, from all sorts of minerals without exception. And thence I may be allowed to conclude that there is not any element that is an ingredient of all bodies, since there are some of which it is not so."

Thinking as radical as this could have little immediate effect on many practising chemists. They could not put aside the old view of things giving up concepts so central to their thinking. Neither could they adopt the alternative view which Boyle presented. This was the wide application of the corpuscular hypothesis—hitherto the province of the physicist and the mathematician—to explain chemical changes. It is probably true to say that this new way of thinking was just as speculative and "unbaconian" as was the doctrine of qualities which Boyle was keen to overthrow, and in this connection it is interesting to note that the corpuscular view made its first great contribution at the beginning of the nineteenth century, with the work of Dalton. In Boyle's time, chemistry could make little use of it, and the contribution it did make was rather indirect. What it did, was to interest physicists and mathematicians in chemistry for the first time. In their eyes chemistry became a physical science divorced from medicine and alchemy.

Boyles application of the corpuscular theory begins with the statement of a number of propositions, and these are phrased in the humble, non-committal form of a man who is trying to be a Baconian philosopher.

"It seems not absurd to conceive that at the first production of mixt bodies, the universal matter whereof they amongst other parts of the universe consisted, was actually divided into little particles of several sizes and shapes variously moved.

"Neither do I see why we may not conceive that she (nature) may produce the bodies accounted mixt out of one another by variously altering and contriving their minute parts, without resolving the matter into any such simple or homogeneous substances (Elements) as are pretended."

The propositions are clearly the product of an uneasy mind, they are not absurd or impossible, but they are not demonstrated.

In proposing them the Baconian Ideal is swept overboard, they are speculative and Boyle is anxious that they should cause no controversy. He is concerned that people might consider him an Epicurian, and he takes the trouble to explain that he has not read the works of Lucretius thoroughly. Many of his views are prefaced in such a way that it would be easy for him to dissociate himself. "I have sometimes thought", "I suspected the principle", "I should moreover declare in general (for I pretend not to be able to do it otherwise)." Boyle must have been conscious that the corpuscular view was as speculative as that of the spagyrist or the peripatetic, and he did not want it to harden into one more body of opinion whose authority rested with the past. Appreciating the atheistic origins of atomism, he was forceful in removing possible criticism, asserting that "the most wise Author of things" directed the motions of the small particles. Equipped with these timid propositions, he presented his view on the changes we see in matter. The particles aggregate, dissemble, unite with those of the agents used in analysis (fire or acids), forming new aggregates. They move, collide, have various shapes, and their activities produce the properties long associated with the presence of elements. In all this he gave no indication that he understood the nature of the materials we now call compounds, elements or mixtures. M. Boas says of his work, "there is no set of corpuscles irretrievably committed to a particular set of characteristics which constitute one and only one substance" (*R. Boyle and 17th Century Chemistry*, p. 97).

As Boyle completed *The Skeptical Chymist* he must have looked hard at the fruits of his effort—a destruction of the doctrine of the elements and the tentative introduction of the corpuscular hypothesis, and he wrote :

"And therefore, if either of the two examined opinions, or any other theory of elements, shall upon rational and experimental grounds be clearly made out to me; 'tis obliging but not irrational, in you to expect, that I shall not be so farr in love with my disquieting doubts, as not to be content to change them for undoubted truths.

"I can yet so little discover what to acquiesce in, that perchance the enquiries of others have scarce been more unsatisfactory to me, than my own have been to myself." (*The Skeptical Chymist*, Everyman edn., p. 230.)

Boyle can have derived little pleasure from playing a destructive role. Alchemists claimed that "death" must anticipate "the resurrection", but there is no doubt that pleasure attaches to the latter. Modern chemistry could only rise on the corpse of the "Doctrine of Qualities", and the resurrection happened towards the end of the eighteenth century. But we may doubt whether Boyle's efforts at the "death" were really necessary—for some chemists the old had already died. A contemporary wrote of the elements.

"The term Principle in Chemistry should not be taken in an entirely precise sense; for the substance so called, are not Principles except from our point of view, being so because we cannot go further in the division of bodies; but everyone knows that these principles are divisible into an infinity of parts,

". . . and as chemistry is a demonstrative science, it cannot accept as a basic things which are not real and capable of demonstration." (Lémery, *Cours de Chymie*, 9th edn., 1701.)

It seems that Lémery found Boyle's warning quite unnecessary, and his claim that *everyone* used the term "principle" as a figure of speech is, if true, very damaging to the significance of Boyle's writing on the subject. The last quotation is "the unkindest cut of all". Not only did Boyle spend a great deal of time doing an unnecessary job, but in Lemery's view he was unscientific in the act!

THE EXPERIMENTALIST

In assessing Boyle's contribution to science, historians have differed over his theoretical writing. Some have considered this aspect of his work the most significant, some have found it lacking in originality and very much overrated. But one thing is agreed—

he was a prolific and original experimenter. Looking through a collection of his works, the epithet "Father of Chemistry" does make an appeal. The volume, quality, and diversity of his investigations is quite remarkable.

Experiments are attempts to answer questions, and Boyle had an inexhaustible supply of these. Together with his intense curiosity, he had considerable personal skill as a laboratory worker, but, probably more important, he had a genius for directing the work of others. We might call him a great *director* of research.

His output is all the more remarkable when seen in the light of his other activities, for society made great demands on his time. Like Kepler or Galileo before him, or Nobel long after, he had a tremendous capacity for work. He often refers to the demands of society, and he leaves us in no doubt that many of his commitments were most frustrating :

"But being oblig'd to make some journeys and Removes, which allowed me no opportunity to prosecute the Experiments, I have made no very great progress."

Or again, explaining the lateness of a publication :

"Your lordship will perhaps wonder, that you have not receiv'd it sooner, as indeed you should have done, if I had been befriended with Accommodations and Leisure."

Writing to his sister, he remarked that in addition to the plague, there were other diseases, one being "fits of the committee", and another "consumption of the purse". The busy life he led certainly interrupted his own practical contribution, but his assistants could carry on in his absence. On the other hand, the task of preparing work for publication could not be delegated so easily, and this suffered more than anything else. It has already been mentioned that the publications of a number of English scientists of the period were fragmentary, lacking in general theme, and that this was due to their regard for science as a fact-finding activity. To appreciate Boyle's work we have to superimpose upon the current English style the confusion brought about by his own activities.

After making allowance for this, it is probably not too harsh to say that Boyle must have been a most unsystematic man. He talks of misplaced manuscripts, apologises for incomplete diagrams, begs our pardon for the repetition of experiments (admitting that he was not sure at the time of going to press whether they had been published previously).

"I made in the year 1660 several observations, that would not perhaps be impertinent in this place, yet having long since left them with a friend, who lives far off, and not having them now in my power, I must beg your Lordships permission to reserve them for a part of the Appendix, which I doubt I shall be engaged to add to this Epistle." (*A Continuation of New Exp.*, 1669, p. 68).

His experiments were always relevant to the question in hand, but his descriptions of them might at any time be side-tracked by any one of his many intense interests. An illustration of a point might develop into a discussion occupying pages, and the reader, finding himself lost, is constantly referring back. If we add together his illustrations, we get the impression of a man with a wide general knowledge, who is particularly interested in geography, history, and languages.

Boyle is popularly regarded as a physicist, because his studies on the physical characteristics of air are common knowledge. An inspection of elementary chemistry textbooks gives little hope that the view will be corrected; Boyle gets little more than a mention in them. Yet if his work is divided on a subject basis, and due regard is paid to the time he must have spent on physical and chemical work, then he emerges as a chemist. He was a chemist who pioneered the use of physical techniques, and, before studying particular researches, it might be useful to list some of them: acids and alkalis; the air; phosphorus; the nature of fire; heat and cold; qualitative analysis; the colour of chemicals; medicines; minerals; the odour of chemicals; nitre; dyes; glass; combustion and calcination (rusting). Contemporary chemists had difficulty in recognizing him as one of their number. After all, he had set himself against the doctrine of the elements; he applied physical methods in chemical study; and he tried to apply the corpuscular hypothesis while admitting its limitations. "For I presume you would desire, as well as I, to learn (at least) why the particles of mercury, of the tartar, and of the acid salts convening together, should make an orange colour than a red, or a blue, or a green." It is harder to understand why we also fail to regard him as a chemist. His attitude to the doctrine of the elements can hardly offend us, we are happy with the use of physical methods, and our own atom leaves much to be desired. The reason may be found in his method of working and writing; there are so few definitive statements to relate to his name.

If Boyle had no large theme to inspire him, then how did he work? How did a question emerge and what was the source of his energy? Most scientists work with a singleness of purpose, they concentrate on the job in hand, and only meaningful observations are recorded; phenomena which appear to be unrelated are unobserved or ignored. For this reason, the ground they cover is worth cultivating by another, and another. Fleming must have seen what many before him had seen, he was different in that what he saw stimulated him. So it was with Boyle. A chance observation in one experiment suggested a question to be answered in the next. He was intensely curious. His knowledge of other branches of science was important. He was quick to see the wide implications of his findings. After describing an experiment which demonstrated the rise of water up tubes packed with fine powders, he was quick to remark that the phenomenon might relate to the rise of water in plants.

"I may take notice, that in the last Tryal above recited, I made Water ascend to near if not above, 3 foot $\frac{1}{2}$; and if by so sleight an Expedient, Water may be made to rise as high as is necessary for the Nutrition of some thousands of Plants, (for such a number there is, that exceed not 3 foot $\frac{1}{2}$ in height), one may without absurdity ask, why tis not possible that Nature, or rather the most wise Author of it, may have made such Contrivances in Plants, as to make Liquors ascend in them to the tops of the tallest Trees" (*A Continuation of New Exp.*, Part 1, 1669, p. 96).

In the same series of experiments he describes the construction of a portable barometer or "baroscope", and this is followed by a long discussion about the height of mountains, together with references to travel and navigation. In this instance, too, his interest in other subjects may have been the inspiration for his practical work. Some of the investigations conducted with his vacuum pump show the span he had. Who else would have tried to study "the slaking of Quicklime in the Exhausted Receiver", "the production of Heat by Attrition friction in the Exhausted Receiver", "the production of a kind of Halo and Colours in . . . .", "the production of light . . .", "the propagation of Sounds . . .", "the falling in the Exhausted Receiver, of a light body," "the attractive virtue of the Loadstone . . .", "the seemingly spontaneous Ascension of Salts along the side of glasses", "height to which pure Mercury and Mercury Amalgam'd with Tin will stand in Barometers", "the bending of a Springy Body in the Exhausted Receiver".

It has been said that important advances in science stem from the application of a new technique, and it is easy to find examples of this. The recent introduction of chromatography made possible the rapid progress in protein studies; the invention of photography opened up new horizons in several sciences. The rapid adoption of new techniques to a wide field of study depends upon good communication between scientists, and upon the specialist being interested in the work of others. Boyle could number mathematicians and physicists among his friends, and there is little doubt that he interested them in chemistry, and that they in turn stimulated many of his physical studies.

Another interesting feature of his work is his attitude to quantitative study. In some of his work he made measurements, recording length, weight, and time very carefully. But it is his occasional indifference to quantitative results which presents the problem and which sets him apart. For example, why was he so inconsistent in his application of gravimetric techniques? This aspect of his work is illustrated in the famous transmutation experiment. Still anxious to destroy the four-element theory he attempted to convert "water" into "earth", and this he did by boiling water in a glass vessel until a sediment appeared. A modern chemist would demonstrate the significance of this by making the study quantitative. The flask would be weighed before and after to eliminate the possibility that this was the origin of the sediment. The purity of the water would be checked to confirm that dissolved solids were not the cause. The experiment would be repeated in metal vessels to confirm that water does not react with glass. The interesting point is that Boyle recognized these various controls and yet failed to exercise them.

"It were also fit to know, whether the glass body, wherein all the distillations are made, do lose of its weight anything near so much as the obtained powder amounts to over and above the decrement of weight, which may be imputed to the action of the heat upon the substance of the glass.

"Which scruple and some of the former I might have prevented, if I had the convenient metalline vessels, wherein to make the distillations instead of glass ones."

It is tempting to explain his attitude in terms of a lack of mathematical appreciation, and to see this demonstrated by the absence of mathematics in his writing. But even Newton could not apply his mathematics to seventeenth-century chemistry, the subject was not in fit condition to receive it. The fact that Boyle failed to demonstrate mathematical prowess simply reflects the state of the subject. In a few places he does demonstrate a knowledge of the subject, and we know that his early education included mathematical studies. He was interested enough to ask Dr. Wallis whether a jet of water from a fountain described a parabola, and, in describing the almost equal rise of mercury in tubes of different diameter, he wrote :

"but in case these had been equal, then the Solidarities of the cylinders mercury would have been to one another as their Bases; and since these, being circular, are in duplicate proportion to their Diameters, that is, as the Squares of their Diameters; its plain, that if the Diameters be as one to two, the squares of them must be as one to four; and these Cylinders consisting of the same Mercury, their weights will have the same Proportions with their Solidarities, and consequently would be as one to four" (*A Continuation of New Exp.*, Part 1, 1669, p. 12).

This kind of writing demonstrates that he had some mathematical ability. It is possible that Boyle wished his work to be accessible to a wide public, and that in his view a rigorous, formal presentation would not serve this end. There is also the fact that quantitative studies make great demands on both time and apparatus, and both of these were in short supply. It would not be easy in the "transmutation experiment" to obtain metal vessels, and there were so many phenomena awaiting investigation. In the same experiment controls would have certainly caused the rejection of the point he was keen to make! It would be too harsh to make this the reason for his failure to exercise care in this case, but it is likely that he had already made up his mind on the subject, and that his usual scepticism was submerged.

But these reasons are no more than immediate. He *could* have found time and apparatus, and he could have sacrificed the General Reader had he thought it worth while. Boyle seems to have doubted the significance of quantitative results; more than this, they probably filled him with suspicion. He was indifferent to magnitude because he did not regard it as invariable. Our interest in measuring a combining ratio stems from a conviction that it is constant, he had no incentive. No doubt he would have been appalled at the statement "For a given mass of gas, at a given temperature, the volume is inversely proportional to the pressure".

Mathematical formulations have general implications and Boyle was too much of a Baconian to use them. He was far too busy collecting stamps to attempt their arrangement in the album.

When he used quantitative techniques his interest was in demonstrating a phenomenon; magnitudes were unimportant. He weighed a sample of silver, converted it to silver nitrate, and weighed the product. The gain in weight was significant; it meant that chemical combination had occurred; but in the absence of a law of constant composition there was no incentive to think more of the actual data. His interest in a sensitive balance was aroused, not because he wanted to weigh accurately, but because using it, he might detect phenomena associated with only slight changes of weight.

On occasions he found it necessary to defend his attitude to experimentation, and he wrote as though he had been criticized for lack of care and precision.

"And let me add on this occasion to what I have elsewhere said to the like purpose, that tis a great discouragement to many ingenious men, and no small hinderance to the advancement of Natural Philosophy, that some nice Criticks are so censorious in exacting from Attempters the very best Contrivances, and many that would be attempters stand to much in awe of such mens judgements; for though in very nice Experiments the exactnesse of instruments is not onely desireable and useful, but in some cases necessary; yet in many others, where the production of a new Phenomenon is the thing aimed at, they are to be looked upon as Benefactors to the History of Nature, that performe the substantial part of a Discovery, though they do it not by the most easie and compendious wayes deviseable, or attain not to the utmost preciseness that might be wished, and is possible." (The Preface, *A Continuation of New Exp.*, Part 1, 1669.)

THE USE OF PHYSICAL TECHNIQUES IN CHEMISTRY

Boyle's application of physical methods to chemical investigations is one of his most significant contributions. The balance was regularly used as a research tool both in the tracing of chemical reactions and in the identification of substances; it was also used for the detection of loss by evaporation. Ironically, some of his quantitative work was responsible for erroneous thinking which lasted for a hundred years, and which provided Lavoisier with the material for some of his most famous experiments. It is interesting that despite his own warnings on the danger of forming conclusions, he himself fell a victim on occasions, and that his own errors of judgement provide excellent proof of the point he so often made. The "transmutation experiment" has been mentioned already, and this was one of the studies corrected by Lavoisier. In another very important series of experiments, Boyle attempted to weigh fire, and this he succeeded in doing. A wide variety of solids, including metals, were placed in glass retorts. The retorts were then warmed to expel air, the necks were sealed, and heating was intensified. Most of the substances gained weight, and Boyle concluded that fire had been trapped and weighed. Lavoisier, a hundred years later, showed that the results were consistent with his new idea of combustion and calcination. This piece of work stands as an excellent illustration of the dangers of interpreting results—a danger which Boyle recognized. The part played by the air was not taken into account. At the beginning of the century, Van Helmont had made a similar mistake in his demonstration that water was the only element. In this, a plant was grown in a pot and was provided with water only. The plant gained weight and Helmont concluded that water had been converted into plant substance. Here again, a conclusion was made in ignorance of a very significant factor—the atmosphere.

But to return to happier experiments in which he blended his physical methods and his Baconian ideal. It has been remarked that Boyle was interested in phenomena, and that measurements only had meaning when they demonstrated a process. This is true, but it does not mean that he was indifferent to sensitive instruments. Some phenomena required sensitive detectors, and when this was the case, Boyle showed himself to be very interested in the performance of his apparatus. Writing on "The Atmospheres of Consistent Bodies" he showed that many solids lose weight on exposure to the air. He showed that ice evaporates, and that over a period of a few hours there is a noticeable change in the weight of a piece of wood. Commenting on the current view that an antimony vessel was capable of poisoning wine, without itself losing weight, he says :

"I will not urge that tis affirm'd, not onely by the generality of our Chymists, but by learned modern Physitians, that when either Glass of Antimony, or Crocus Metallorum impregnate wine with Vomative and Purgative Particles, they do it without any decrement of their weight; because the scales in Apothecaries Shops, and the little accurateness wont to be imployed in weighing things, by those not vers'd in Statical Affairs, makes me (though not deny the Tradition which may perchance be true, yet) unwilling to build upon observations." (*A Continuation of New Exp.*, p. 195.)

In the famous "transmutation experiment", Boyle suggests that a sediment might be identified as glass by measuring its specific gravity. The analysis of alloys in this way was a technique employed by craftsmen long before Boyle, but the general application to chemistry must be credited to him. Discussing the analysis of mineral waters, he provided a most extensive routine, including tests for sulphur, iron, copper, acidity, and alkalinity, and he included a measurement of the specific gravity of the sample. His excellent study of the properties of phosphorus included details of its physical characteristics, including its specific gravity. Boyle's work with the vacuum pump included a study of the heights to which salt water would rise in a partially evacuated tube, and he saw the possibility of measuring the salinity of different samples by comparing their heights under the same conditions. He also discussed the possibility of using the same technique for the analysis of vinegar and alcohol–water systems. Elsewhere he describes the importance of temperature in this connection, and he notes that a variation can render the technique unsatisfactory.

THE NOTION OF PURITY

Impurities give rise to a great deal of confusion in the interpretation of chemical reactions. Their presence can give rise to observations which are not easily reproduced, and the realization of this was one of the major advances in chemistry. Without such an appreciation, there can be no concept of a chemical compound in the modern sense, because no two samples behave in exactly the same way. For us, copper sulphate is a blue crystalline solid which is soluble in water, which gives a black precipitate with hydrogen sulphide, turns deep blue on the addition of ammonia, etc. If the particular sample we are using is not completely soluble in water, we are quick to suggest the presence of a small quantity of another substance–an impurity. Without the idea of "impurity" we would have to suppose that there were many kinds of copper sulphate, each one being slightly different from the rest.

One of Boyle's greatest contributions to the subject was his study of non-reproducible results, and their causes. He was convinced that one of the main troubles lay in impure solvents or "menstruums". Water and mineral acids should be carefully tested to detect impurities, and chemists should discriminate between their sources of supply. His method for trace analysis was one of his most systematic contributions and it set the scene for later quantitative work.

THE USE OF ASSISTANTS

It has been remarked that Sir Humphry Davy's greatest contribution to science was not the invention of the safety lamp, or the discovery of sodium, but was the appointment of Michael Faraday to a post at the Royal Institution. In the early nineteenth century, courses of instruction in science were rare in England, and anyone keen to learn had to seek employment with a practising scientist. So it was that Faraday learned by direct contact with Davy. One and a half centuries before this, Boyle was responsible for the scientific education of his assistants. Men were employed to help in the working of the laboratory; their education was a derivative, but a most important derivative of his method of work. Assistants were carefully employed according to their abilities. Some were supervised at every stage of the work, while others had the free range of a wide topic. Robert Hooke worked for Boyle, and the experience was probably responsible for his interest in chemistry. Denis Papin, well known for his later work on steam-engines, assisted Boyle with his pneumatic experiments. Homberg assisted for a short time, and it seems likely that his later work on acids and alkalis was inspired by the experience. The terms of employment of the assistants varied enormously. Some were financially dependent on Boyle. Some were men of independent means who had already established themselves as scientists and who were paying courtesy visits. They would come knowing what work was in progress, and leave to employ the ideas and techniques in their own laboratories. Nothing seems to have been withheld from these visitors. Boyle was most generous with his knowledge and skills, and entertained no desire for personal gain. The history of his research laboratory is that of any other; at first he had difficulty in getting the right assistance, but later, when his reputation was established, he had the embarrassment of too many volunteers.

THE FOUNDATION OF THE ROYAL SOCIETY

The establishment of scientific societies was a most important aspect of the Scientific Revolution. They derived from an increasing interest in science, and in turn they added to its progress. Regular meetings of scientists allowed an all-important cross-fertilization of ideas. Before their formation, the individual worker communicated with a close circle of friends by exchanging letters. The wider circulation of his work had to await the publication of a book, and this might take years to compile; meanwhile the individual experiment or idea was withheld. The scientific societies accelerated progress by publishing fragments, so that a scientist could make his contribution to another's research while it was in progress. The list which follows gives the dates and origins of some of these early societies, and the foundation of the Royal Society may be seen against a general trend.

1560 Naples. Accademia Secretorum Natural
1603 Rome. Accademia dei Lincei

1651 Florence. Accademia del Cimento (lasted 10 years)
1662 London. The Royal Society
1666 Paris. Académie des Sciences
1700 Berlin
1725 St. Petersburg
1741 Swedish Accademy
1743 Danish Accademy
1743 American Philosophical Society

The origins of the Royal Society are to be found some 20 years before its emergence, and two roots may be distinguished. In 1645 the Gresham Group drew up a formal constitution. According to John Wallis, mathematician and priest, the group met weekly, paying weekly contributions to be used for the cost of experiments, and they discussed the "New" or "Experimental Philosophy". The discussion of theological or political matters was expressly forbidden. Wallis gave a list of the early members of the group, and Boyle's name is absent. An 18-year-old recently returned from Europe would be quite unknown to the members.

On Boyle's return to London he had a short stay with his sister Katherine before going on to Stalbridge, and through her he came in contact with a second group of men who expressed an interest in science. These were the Comenians, disciples of the Czech educator Comenius, who fostered the view that a universal language was necessary in order to achieve universal understanding and its derivative—universal peace and goodwill. Samuel Hartlib was the most active member of the group. There was no formal constitution or regular meeting place. Members kept in contact by letter, and Hartlib acted as intermediary in most discussions. Boyle named it the "Invisible College", an apt enough description in view of the loose structure. During his stay at Stalbridge he was a very active member. As a vehicle for advancing science it was far inferior to the Gresham Group and, but for Boyle's membership, it would be unimportant in a discussion of scientific societies. It most serious defect was the lack of close contact between the members; nevertheless, it was responsible for encouraging Boyle in his early scientific work. Amongst its members may be listed John Dury, John Pell (mathematician), Theodore Haak, Benjamin Worsley (surgeon), William Petty, Frederick Clodius, and Robert Child and George Starkey (alchemists). Writing to a friend at Cambridge, Boyle expresses his delight at being one of the group :

"The best on't is, that the cornerstone of the Invisible, or (as they term themselves) the Philosophical College, do now and then honour me with their company . . . men of so capacious and searching spirits, that school philosophy is but the lowest region of their knowledge."

Though Hartlib had no experience of science, he tried to communicate technical information to members who expressed an interest. In 1647 he wrote to Boyle on the subject of pneumatics, and Boyle's reply shows that he was well abreast of developments in this field; at the same time there is the clear indication that he was far from being a purely receptive member; he was already in a position to advance knowledge. During this period he showed himself to be very interested in medicine, a subject which had special significance for all Comenians because it had an immediate contribution to make to man's estate. Boyle's interest in the subject intensified and grew more morbid as he got older. He was preoccupied with his own health, amassing a large collection of cures from many sources, and it has been suggested that his interest in chemistry sprang from a concern about medicines.

In 1652 Boyle visited Ireland, and this may be regarded as the start of his movement from the "Invisible College" towards the Gresham Group. Even before this time there had been some contact between the two. Haak had become a member of both, and Merret of the Gresham Group knew Robert Child. Some members of the Gresham Group had moved to Oxford by this time; their sympathy with the Parliamentary Cause had led to various appointments at the University. Wilkins had become Warden of Wadham, and Goddard Warden of Merton; Wallis was Savilian Professor of Geometry. William Petty, already mentioned as belonging to the "Invisible College", associated with them while he was at Oxford, and afterwards took up a political appointment in Ireland. Boyle and Petty met in Ireland, and no doubt the Oxford (Gresham) Group would be discussed. It was to Wilkins that Boyle turned when he was anxious to get a laboratory assistant, and Wilkins's reply to the request is interesting because it shows the relationship between Boyle and the Gresham Group at this time (1653):

"If it be not, Sir, prejudicial to your other affairs, I should exceedingly rejoice in your being stayed in England this Winter, and the advantage of your conversation at Oxford, where you will be a means to quicken and direct us in our enquiries. And though a person so well accomplished as yourself, cannot expect to learn anything amongst pedants, you will here meet divers persons, who will truly love and honour you."

The style of the letter is formal, and it is clear that the two men were strangers at this time.

It is clear that as time went on the "Invisible College" failed to satisfy Boyle's interest in "experimental philosophy", and that he turned to the Gresham–Oxford Group because their contribution to the subject was so much greater.

In 1654 Boyle moved to Oxford, and this became the centre of his activities for the next 14 years. During this period he did some of his most important work and, not least, he succeeded in interesting a number of the most able physicists and mathematicians in chemistry. The work he did gained international recognition, so that when the Gresham–Oxford group reformed in London in 1660, and could name Boyle as one of their number, his reputation became theirs. To the value of his scientific reputation must be added the importance of his social position. He alone was in a position to be received at Court, and his influence was a significant factor in the elevation of the group to the title of "The Royal Society" in 1662. Among his other contributions to the success of the early Royal Society may be mentioned the appointment of its first secretary and demonstrator. Henry Oldenburg was appointed to the first post and Robert Hooke to the second. Oldenburg had been tutor to Boyle's nephew, and he had been in contact with Boyle over scientific matters for a number of years. Hooke had been with Boyle in Oxford, and his merits as a technician had already won him a high reputation. In his new post he emerged as an independent thinker, and the publication of his *Micrographia* was one of the major events in the Society's early history.

True to the Baconian Ideal, the early members of the Royal Society met to discuss problems from any branch of science. Wren, the creator of St. Pauls, would discuss chemistry with Hooke and mathematics with Willis. Thanks to Boyle, chemistry was one of the major topics of study from the beginning. In Paris, the Académie was conducted in an entirely different spirit; there was a much stronger element of professionalism. If chemistry was discussed, the participants would be chemists, medicals, and botanists, and a man like Huygens would make no contribution at all. The influence of a Boyle was absent. It is interesting that Homberg became a member of the Académie in 1691, and that chemistry came into prominence soon afterwards. Sufficient to say that Homberg had previously spent some time working in Boyle's laboratory.

In his later years, Boyle's interest in the administration of the Society seemed to wane, and so, too, did his enthusiasm for some of its work. In 1680 he was chosen President, but he declined to take office and Wren received the appointment. A short time afterwards he refused to assist the Society in their study of the nature of alcohol, preferring, instead, to censure them in the best *Sceptical Chymist* tradition. It was as though he anticipated his own epithet the "Father of Chemistry".

FURTHER READING AND REFERENCE

Boas, M., *Robert Boyle and Seventeenth-century Chemistry.*
Boyle, R., *The Skeptical Chymist* is widely available (e.g. Everyman edn.).

CHAPTER 2

MARCELLO MALPIGHI, 1628–1694

J. S. WILKIE

INTRODUCTION

IN THE seventeenth century there was undoubtedly a reawakening of what are now called the "biological sciences" : but was there a revolution? Whatever the younger workers of that age may have seemed to their older contemporaries to be, it is now very difficult to view them as revolutionaries. Indeed, it is for us somewhat easier to discover and to specify the traditional elements in their work than it is to say in exactly what way their ideas and methods were novel.

In attempting a general estimate of the progress made in the biological sciences in the seventeenth century, it is natural to begin with a brief consideration of the work of Harvey, and this is particularly appropriate when our principal concern is an account of the contributions made by Marcello Malpighi. Malpighi was born in the year in which Harvey published his most important work, the *Exercitatio de motu cordis et sanguinis*; this is a trivial fact but it helps us to realize and to remember that Malpighi belongs to the generation immediately following that of Harvey.

To appreciate the full novelty of Harvey's *De Motu cordis* we should have to exhibit in detail how the heart's motion was discussed by the anatomists whose works were used as textbooks by Harvey when he was a young man. This would present no insuperable difficulties for there is no doubt at all that Harvey studied principally Caspar Bauhin's *Theatrum anatomicum* of 1605 and Du Laurens's *Historia anatomica humani corporis* of 1600.

It is, however, obviously impossible to present here long extracts from either of these books. It might, indeed, well be asked why such lengthy quotations should be desirable, even were there space for them. They would be desirable because they would illustrate, to anyone even superficially acquainted with the *De Motu cordis*, the difference in spirit between Harvey and his immediate predecessors. Their treatment of the subject is hesitant and confused, and they rely principally upon authorities and very little upon first-hand observation except for the facts of mere dead anatomy. Indeed, if we had nothing more to go upon than a comparison of Harvey's work with that of his immediate predecessors, we might well be tempted to consider Harvey a revolutionary. In fact, however, we cannot be contented with such a comparison, for Harvey justly supposed that he had learned much from a direct study of the writings of Aristotle; and if he seems to speak slightingly of Galen, this is only in reaction against the Galenists, that is those who slavishly followed the letter of the Galenic texts without being moved by their spirit. If we compare Harvey, not with Bauhin or Du Laurens, but with Galen, we receive at once an impression of continuity; Harvey begins, one might say, where Galen left off. It is as though Harvey belonged to the generation next after Galen's; and if we consider not only the *De Motu cordis* (which is a work of maturity, published when Harvey was 50) but also Harvey's earlier writings, we find it difficult to say whether he should be considered the first modern or the last Greek physiologist. Nearly all the facts used by Harvey to establish his case, including the anatomy of the valves of the heart and the very important anatomy of the foetal blood system, were known to Galen, and there is very little that is new in Harvey's methods. Galen was well aware of the nature and even of the logic of experiment, and he more than once used ligatures to demonstrate direction of flow in vessels; his principal defects in his treatment of the heart and blood-vessels were his inability to appreciate the importance of exact quantitative considerations and his inability to rid himself totally of a theory (that of Erasistratos) which he had himself shown to be untenable. Harvey's use of measurement (of the capacity of the heart) is certainly an innovation of fundamental importance, but there is little evidence that Harvey fully appreciated either its novelty or its importance.

This digression is, I think, necessary if we are to appreciate what was and what was not novel in Malpighi's methods : novel, that is, particularly when compared with the Galenists, who were his most pertinacious critics : though indeed his methods were in some interesting respects in advance also of those used by Harvey. Malpighi was undoubtedly judged by many of his contemporaries to be a rather dangerous innovator, and he also thought of himself as having made a fundamental break with the past; for he tells us in his autobiography that, as a young man in Bologna he studied the peripatetic philosophy "for several years" beginning in 1645, but having removed to Pisa in 1656 he was instructed by Borelli in the *libera philosophia.* What he meant by this expression seems to be the methodological doctrines of Galileo, with which Borelli, who was a mathematician and astronomer as well as a physiologist, was well acquainted. In his writings Malpighi does not, in fact, show any great interest in philosophy, and the new point of view which he adopted, and which he seems to have thought of as a new *philosophy*, appears to have been simply a resolution to study nature directly by observation and experiment and not indirectly through the writings of others. Thus, in his introduction to his treatise on the kidneys, he writes :

"Do not stop to question whether these ideas are new or old, but ask, more properly, whether they harmonize with Nature. And be assured of this one thing, that I never reached my idea of the structure of the kidney by the aid of books, but by long, patient, and varied use of the microscope. I have gotten the rest by the deductions of reason, slowly, and with an open mind, as is my custom." *

Malpighi, however, was not so foolish as to pay no attention at

* The translation is due to J. M. Hayman, Jr., *Ann. med. Hist.* **7**, 242–63 (1925). I am not sure whether *animoque remisso* really means "with an open mind", but in the context this rendering seems extremely plausible.

all to the work of earlier scientists : he simply wished to control their observations by his own, and to make further discoveries for himself. This is entirely in the spirit of Galen, though not of the Galenists who harassed Malpighi.

If we compare Malpighi's writings with those of Harvey, considering all Harvey's works and not simply the *De Motu cordis*, we find a very striking difference; for Harvey is constantly borrowing from Greek sources, principally from Aristotle. Moreover, in his earlier work he uses arguments which have no meaning outside the complex of ideas peculiar to Greek biology, and his last book, the *Exercitationes de generatione animalium* of 1651, is the work of a devout Aristotelian.

Malpighi usually refers to Greek sources only in the most summary and general way at the beginning of brief reviews of the opinions of earlier writers; and he very seldom appeals to any Greek author as to an authority : the significant exception being the Hippocratic writings. An explicit and conscious use of Aristotelian theoretical notions occurs, I think, nowhere in his works.

Thus between Harvey and Malpighi there is, as I have said, a striking difference. It would, however, be a great mistake to suppose that the absence in Malpighi's writings of explicit references to Greek authorities and ideas constitutes evidence for a revolutionary change of method. If we compare his work with what is best in Galen, or with Harvey's *De Motu cordis*, we find little that is novel in this respect. Sometimes, indeed, as in his treatise on the liver, Malpighi resembles Galen rather more than he resembles the Harvey of the *De Motu cordis*; for we become uncomfortably aware that a great deal of argument is based upon extremely few reliable observations; some of the conclusions, indeed, may be sound enough, but they are very weakly supported. Galen's excessive addiction to teleology is certainly not imitated by Malpighi, but this is a matter of degree, not a question of principle; for Malpighi, like all other physiologists, is, of course, interested in the question of what a liver is *for*.

However, there are some points of methodological interest in

Malpighi's works. He is able to profit by corpuscular or atomic notions of which Harvey is demonstrably ignorant, and his use of inductive arguments in his study of the respiratory organs of insects is something which, I think, is not to be found, except in an extremely rudimentary form, in the Greek authors, and which I do not remember having encountered in Harvey's writings. I shall deal briefly in this place with Malpighi's use of a corpuscular hypothesis, and shall consider later on his special use of induction (see below, "Malpighi as Experimentalist").

When we speak of "method" we now commonly mean arguments of a particular logical form, or some special set of explanatory concepts. In this sense it would be improper to refer to Malpighi's use of the microscope when we are considering his methods. However, there is a sense in which the use of the microscope does constitute a difference of formal method, as between Harvey and Malpighi. For Harvey not only failed to make use of the notion of atoms or corpuscles in principle unobservable, he also suffered from a curious insensitivity in respect of objects invisible to the naked eye though in principle observable, and this defect deserves to be considered as one belonging to formal method, or, at least, as lying in the fringe of formal method. Thus, in his *Exercitationes de generatione animalium* he describes how he dissected a doe immediately after coition and found "absolutely nothing" in the uterus. He says that the huntsmen who were present were greatly surprised by this, but were convinced of the fact when he persuaded them to examine the organ "with their own eyes". We are forced to conclude from this that Harvey, who seems sometimes to have used lenses of low magnification, did not simply reject the microscope of his time as technically inadequate but was really unaware of the necessity of considering the existence of entities of microscopic dimensions.

Let us now consider the difference between Harvey's extremely confused notions of solution and the clear and precise model employed by Malpighi. Harvey could not fail to possess some idea of solution, and that he did so is shown by his writing, in the Introduction to the *De Motu cordis*: "No one denies that blood as such . . . is imbued with spirits" and "blood and spirits constitute one body, like whey and butter in milk, or heat and water in hot water." * But that his notion of solution was troublesomely confused is shown by his asserting (in the same Introduction) that air could not have entered the blood in the lungs because no air could be *seen* in the blood of the pulmonary veins after forcible distension of the lungs of a dog. He writes also :

"Or do seals, whales, dolphins and the whole cetaceous tribe, and fish of every description, living in the depths of the sea, take in and emit air . . . through the infinite mass of waters? For to say that they absorb the air that is infixed in the water, and emit their fumes into this medium, were to utter something like a mere figment."

Malpighi, on the other hand, gives, in his first letter on the lungs (1660), a very clear theory of solutions.

"From what I shall subjoin", he writes, "I shall believe probable that the lungs were contrived by Nature for mixing the mass of the blood. . . . And, in the first place, there is no need for me to persuade you by a long string of words that there exist in Nature some bodies which have not received a fluidity proper to themselves, but which have even their least parts so prepared for easy joining and union that they cannot be disjoined except by the greatest force, and when once disjoined tend to mutual union; these same bodies are with great ease made fluid by mixture with another body interposed; this we see firstly in metals and other *bodies* fused by fire, then in tartareous parts which, fused and mixed with water are fluid and invisible, but when united compose the solidity of stone; again by *aquae regiae*, acids and the like, being interposed and occasioning discontinuity, we cause the most solid metals to become fluid; nay, indeed, what is more wonderful, flow occurs from mere discontinuity of parts even in the driest *materials*, as in burnt lead and tin when ground up in a mill by the potter; for such is the grossness of our sense that they readily believe the smallest grains, even of sand, to flow like water."

* Translations by Willis, Sydenham Society, 1847.

This passage, like almost all passages of Malpighi's Latin, contains some linguistic difficulties, but these cannot be discussed here. The reference to fusion of metals is pointless unless we suppose, what is not improbable, that Malpighi believed fire to be corpuscular. "Burnt lead and tin" are, I believe, what we should call oxides.

That a clear theory is here expressed seems to me to be shown by the insistence upon dissolution of continuity and interposition of particles of some other substance.

We must, however, note that a clear basic theory does not enable Malpighi to find an acceptable theory of the function of the lungs. The theory of the function of the lungs that he gives is based on the discovery of the lymphatic system by Aselli (1627), Pecquet (1651), and Bartolin (1653). Malpighi supposes that the "white blood", as he calls it, adding that it is the "lymph of Bartolin", which enters the venous system near the heart, must be mixed with the "red blood" returning in the veins, and that it is the function of the lungs to perform this mixing. The weakness of the theory, from a contemporary point of view, is that Malpighi cannot show that sufficient mixing would not occur in the heart itself.

It was to provide an experimental refutation of this theory that Hooke (*Phil. Trans.*, 1667) devised one of his most ingenious experiments: the lungs of a living dog were kept constantly expanded by a stream of air provided by two bellows working in series, the air escaped from the lungs through a number of fine holes made in the lungs by means of a needle. Hooke found that the dog showed no special symptoms while the air-stream was maintained, but exhibited a reversible coma if the stream was interrupted. This experiment shows that no ill effects follow (within the time of the experiment) upon interruption of the churning movements of the lung.

MALPIGHI AS ANATOMIST: 1

It is often said that comparative anatomy began in the late eighteenth and early nineteenth centuries. This statement appears puzzling when we consider that Greek anatomy was in some important sense comparative.

Harvey, it is true, complained that his contemporaries knew only the anatomy of the human body, but the ignorance of which he complained was partly an accident of the close association of anatomy with medicine and partly a peculiarity of the sixteenth century : a consequence of Vesalius's demonstration that Galen had too readily assumed that what he found in other animals would be represented in man. This demonstration of Galen's mistake led some anatomists to despise, or at least to neglect, comparative studies; and, as a result, to treat anatomy as a subject divorced from experimental physiology, since the kind of experiments they were able to perform virtually all involved radical vivisections, and these could not be performed on human beings.

The best anatomists of the seventeenth century, however, following Harvey's example, made great use of comparative anatomical studies of a certain kind. Moreover, the introduction of tolerable microscopes about the middle of the century led to an interest in the anatomy of insects and of other small animals, which could then be profitably studied for the first time.

How then could anyone be led to assert that comparative anatomy did not exist before the late eighteenth century? The solution of this problem lies in a distinction already made by Aristotle, but insufficiently appreciated in the seventeenth century: the distinction between *similar* and *analogous* organs; or, as we now say, between *homologous* and *analogous* organs.

The hearts of vertebrates are similar, though not necessarily identically similar, to one another in structure and in function, but they resemble the hearts of invertebrates only in that they have much the same function. We now say that the heart of a cuttle-fish is an analogue of the human heart, and in saying this we use virtually the same term as that used by Aristotle, who employed in this context the Greek word *analogon*. This word already looks like a technical term in Aristotle's writings, and it is used by Harvey in a clearly limited and technical sense. Unfortunately the words used by Aristotle to characterize the more complete similarity (that of both function and structure) do not seem like technicalities, and the word we use in this sense, *homologue*, was not introduced until the mid-nineteenth century. Moreover, it was only in the early nineteenth century that the defining characteristics of homology, the "laws" of homology, began to be studied for their own sake; and only after this study had been fully developed could the extent of the difference between the organs of one great group of animals and those of another be fully appreciated.

In the absence of this appreciation the more progressive anatomists of the seventeenth century were a good deal too free in applying the results of observations made upon organs of one group of animals to the interpretation of organs of a totally different group, and this tendency is clearly seen in the work of Malpighi. In considering this aspect of Malpighi's work we must distinguish between the heuristic (the exploratory) and the critical use of comparative studies of the kind in which he engaged. The distinction will best be understood by considering an example. In his treatise on the liver he starts his exposition with a consideration of the "liver" of the snail and of the "liver" (what we now call the "Malpighian tubules") of insects. Now if his intention were merely to use the information acquired in the study of these invertebrates to help himself to form a guess as to the structure and function of the liver of the higher animals, the guess so formed being then used as an hypothesis to be refuted or confirmed by direct study of the vertebrate's liver, he would be using the study of the lower forms heuristically; and this would be an instance of heuristic comparative anatomy. This procedure would be entirely legitimate, for whence or how hypotheses are derived is of no significance; but he does not seem to be using his study of lower forms in this way; he seems to assume that, because the "liver" of the snail can be seen to be a gland of external secretion, it somehow follows with at least very high probability that the liver of vertebrates will be found to be a gland of the same general kind. This is an example of what I have called the "critical" use of comparative anatomy : I call it "critical" because it is as though he were using the observations on lower forms to confirm an hypothesis suggested by a preliminary study of the higher, and I have little doubt that this is what Malpighi is in fact doing. However, whether he had first made his hypothesis as to the structure and function of the vertebrate liver, or whether he really did use the study of the snail and of insects to suggest the hypothesis, is of no consequence; what is certain is that, since his belief (that the organs seen in the lower forms are of the same kind as those observed in the higher) has no evidence of any kind to support it, observations on the lower forms cannot be used in any way whatever to *support*, though they could be used to *suggest*, suppositions concerning the higher. I think there is no doubt that Malpighi was in this instance attempting to draw some inference from invertebrate to vertebrate organs, and was hence employing an illegitimate form of argument, and I think he did so because he did not understand the nature and limitations, or "laws", of comparative anatomy. This, I think, is of interest because it lends some support to the belief that comparative anatomy, as now understood, did not exist in the seventeenth century.

It might be thought that, though the method was faulty, the conclusion drawn by Malpighi was correct; namely, that the liver is a gland the function of which is to secrete bile; but he goes beyond this by asserting that this is the only function of the liver, which is unqestionably incorrect. It is probable that this mistake may be largely due to the faulty argumentation arising from the incorrect use of comparative anatomical studies.

Undoubtedly Malpighi had at his disposal evidence suggesting that the liver is not a simple gland like the parotid, with which he compares it; for the blood-supply of the liver is unique and suggests an unique function of the organ. Indeed, the ancient theory of the liver, based upon observation of its peculiar blood-supply, is not excessively erroneous. This theory postulated that the liver converted digested food (chyle) into blood. We now think of blood as a tissue and say that blood is formed where its cells are formed; but the Greeks thought of blood as imperfectly elaborated food; and if this is what the word "blood" means, it is not very far from the truth to say that the liver performs a part of the elaboration of digested food into blood. Even if Malpighi supposed that all the digested food (chyle) passes from the intestine into the lymphatics, not into the blood-vessels, there would still remain for him the problem of why the liver, if it is a gland like the parotid, has such a very peculiar blood-supply; and this problem he ignores.

Malpighi, we may say, failed to handle comparative studies correctly in this very complex situation, but, as I have tried to indicate, it was hardly possible for anyone in the seventeenth century to understand the pitfalls inherent in an uncritical use of comparative anatomical studies. Comparative studies, however, lose many of their dangers when the comparisons are made between animals tolerably closely related, and Malpighi was able to use such studies to great advantage. Thus, as is well known, he attempted to discover in the lung of a dog how the arteries were connected with the veins, and being unable to see the connections in this animal he examined the lung of a frog, in which he saw the capillaries. Some of his critics, however, were as over-cautious in the use of comparative studies as Malpighi himself was over-confident, and, as he tells us in his autobiography, they doubted whether the organ observed by him in the frog could properly be compared to the lung of a dog, since the dog breathes by movements of the diaphragm and ribs, neither of which are present in the frog. Faced with such a criticism the more conservative anatomists of the time would have resorted to long and inconclusive argumentation, but Malpighi approached the problem by direct observation, and was able to show that, though air is not drawn into the lungs of a frog, as it is into those of the dog, yet they *are* filled and emptied of air rhythmically, by pumping action of the floor of the mouth and pharynx.

MALPIGHI AS ANATOMIST: 2

Malpighi's greatest contribution to positive knowledge lay in the field of micro-anatomy. Some indication of the importance of this contribution and of the reliability of Malpighi's observations is to be found in the frequent occurrence of his name in denominations of microscopic parts of animals. Thus there are *Malpighian bodies* in the spleen, the skin has a *Malpighian layer*; there are *Malpighian bodies* also in the kidney, and the gut of insects is furnished with *Malpighian tubules.* It is a mere accident that we do not speak of "Malpighian cells of the cerebral cortex", for Malpighi certainly observed the larger neurone bodies of the cortex together with their axons. Not only was Malpighi the first micro-anatomist of the animal body, but he also wrote the first book on the micro-anatomy of plants, for his book on this subject was published in 1675, the equally famous work of Nehemiah Grew, the *Anatomy of Plants*, appearing some years later, in 1682. Charles Singer tells us that "His treatise on the silkworm is the first monograph on an invertebrate". This work, *Dissertatio epistolica de bombyce*, 1669, contains an excellent anatomy of this small animal and includes the first account of the tracheal system of an insect. Malpighi not only observed and described this system for the first time, but also showed experimentally that it was through the *tracheae* that insects breathe.

He also published the first account of the early stages of the development of the chick, as studied with the microscope.

I wish to deal in this place with Malpighi's contributions to the micro-anatomy of vertebrates, reserving for later sections of this chapter consideration of his other microscopical observations.

We are obliged to call Malpighi's investigations into the microscopic structure of the organs of the higher animals "micro-anatomy", for to speak of this work as "histology" would be anacronistic. Histology, as now understood, is based upon the notion of the cell : it is the study of the way tissues are constructed by cells, consequently there could be no histology before the idea of the cell was elaborated during the first half of the nineteenth century. Nevertheless, Malpighi frequently observed what we now call "cells", and, as we shall see, he realized that they were the most active elements of plants; and there is some absurdity in the importance frequently attached to Hooke's somewhat meagre observations (1665) simply because he happened to use in his description the word "cell". Neither Hooke nor Malpighi, however, can without gross anacronism be said to have "seen cells". When Malpighi observed what we now call "the cell-bodies of the neurones of the cortex" he believed he was looking at glands, and thought that what we call "the axons" were the ducts of the glands. This fact does not, of course, detract from the very great merits of his having been able to observe the actual structures, and having done so with the apparatus available to him.

How exactly he did make his many very remarkable observations it is not at all easy to say. He used a compound microscope of the type used by Hooke, of which a drawing may be found in Charles Singer's *A History of Biology*, revised edn., 1950, p 149. He must also have used an apparatus for concentrating the light of a lamp, such as the apparatus shown in the drawing reproduced by Singer. Hooke's microscope, however, has no substage and no mirror for illumination by transmitted light, and it seems to me that most of Malpighi's observations of structures in the parts of animals must have been made with direct illumination. On the other hand, his drawings of sections of plant-stems must surely have been made with transmitted light. For the observation of capillaries in the frog's lung he recommends two procedures of which the first certainly involves the use of transmitted light. In describing the first method he says that the observation is to be made against the "horizontal" rays of the sun. Whether he used a mirror to obtain a horizontal beam he does not say; but he says explicitly that he used a simple lens (*unius lentis perscruteris microscopio*); that is, he prescribes the use of a single lens, but he is obviously prescribing a method he used himself.

The second procedure is designed for use with the compound microscope (*microscopium duarum lentium*). The lung is to be placed on a glass plate with the light below it. This is the only passage I have found in which transmitted light is referred to. In general he does not say what method of illumination he used. That he gives detailed and explicit instructions for using transmitted light in examining the lungs suggests that this form of illumination seemed novel to him.

I do not think it possible that he should have been able to make microscope sections of animal material, sections, that is, which could be examined with transmitted light. The word *sectio* which has sometimes been translated "section" (as in Hayman's translation of the treatise on the kidney) usually means "dissection", and we should need special evidence to prove that in any particular passage Malpighi used it with the meaning of "section". In the plant anatomy, where the drawings are clearly of sections, Malpighi uses in his description not the noun *sectio* but the past participle *secta* (*Portulaca . . . per transversum secta*). We may note that, in discussing the section of the stem of *Portulaca* he says of the cortex *variis utriculorem seriebus constat*. This is unquestionably a reference to what were in fact living cells.

In a passage in which he tells us how he observed the "glands of the cortex" (the cell bodies of the larger neurones), he says that he lightly boiled the brain, we should say he fixed it with boiling water. He says that the "glands" can be seen, though with difficulty, in the fresh (unfixed) brain "because they are pellucid and colourless". This might mean that he observed a *section* of the fresh brain, but it is much more likely that he was merely looking at a small fragment, possibly teased with needles. He says:

"But they are easier to see in the cooked brain, because their substance, which becomes dense on cooking, makes the spaces between them easier to see. . . . By pouring on ink, and then lightly wiping with cotton, the limits of these glands can be clearly perceived; for the intervening spaces are blackened, so that the glands show up clearly against them."*

This seems to show that he was observing a cut surface of the cortex (probably the surface of a section of some millimetres in thickness) under direct lighting. Had he been able to cut a section

* Those who have recently denied that Malpighi actually saw the cells of the cortex seem to base their arguments on the supposition that it was the cells ("glands") he stained, which is clearly false.

so thin as to be translucent and suitable for examination with transmitted light he could hardly have wiped it with cotton-wool. Moreover, he describes no technique for rendering fixed materials translucent; that is, for clearing them.

He speaks of breaking and tearing the materials under examination, and it seems likely that some of his most remarkable observations were made by methods having the effect of teasing. I suspect that he saw the "Malpighian bodies" in the kidney by making slices of the organ and then tearing them across, observing fine structures projecting from the torn edges. However, he says explicitly that he injected the blood-vessels with coloured fluids, and this might have enabled him to see the "Malpighian bodies" in relatively thick sections of the kidney, especially if the tissues were still alive and comparatively translucent.

I believe that the microscopists of the eighteenth century were greatly inferior to Malpighi, and that many of his observations were not confirmed until the nineteenth century; the microscopes of the eighteenth century were inferior in resolution to those of the seventeenth. I may quote here what was written by Sachs with reference to Malpighi's microscopical observations on plants :

"These two works of Malpighi and Grew, so important not only for botany but for the whole range of natural science, were not followed during the course of the next hundred and twenty years by a single production, which can claim in any respect to be of equal rank with them; that long time was a period not of progress but of steady retrogression." (Julius von Sachs, *History of Botany*, Oxford, 1906, p. 244.)

MALPIGHI AS EXPERIMENTALIST

Malpighi was primarily an observer, and does not record a large number of experiments; but those he describes are not without interest.

In his treatise on the liver he records a simple but conclusive experiment in refutation of the hypothesis that bile is secreted in the gall-bladder itself and not in the liver. He writes :

"To put an end to this controversy and to demonstrate the natural pathway of the bile, I conducted the following easily performed experiment, which, however, did no more than confirm my expectations.

"In a cat a few months old, in which the gall-bladder was unusually prominent, I tied off the neck of this bladder with a thread and emptied it through a wound; I then at once closed with a ligature the end of the common bile-duct, where it opens into the intestine; later on, the animal remaining alive for a very considerable time, I found the bile duct which had been intercepted was swollen, together with the common duct. Moreover, that I might remove all probability of any activity of the bladder in the secretion of bile, having firmly ligatured its neck I cut off the bladder itself and totally removed it, and still I found swelling to ensue in the ducts already referred to, as a result of the bile flowing into them; moreover, with my finger I attempted to force the bile upwards in the ducts swollen in this way, and it immediately returned with a rush, and could only be restrained with considerable force."

By the expression "the bile duct which had been intercepted" he means, I think, the duct above the gall-bladder, which had been intercepted in the sense that its connection with the gall-bladder was tied off.

This experiment is in every way comparable with Galen's experimental demonstration that urine is not secreted in the urinary bladder itself. Galen's experiment is somewhat more elaborate, but his use of ligatures is exactly the same as Malpighi's; and just as Malpighi infers that the bile must come from the liver, if it does not come through the walls of the gall-bladder, so Galen infers that urine must come from the kidneys. So far as this simple experiment is concerned, therefore, Malpighi's method is one already used by Galen. Elsewhere, however, Malpighi uses a method that seems not to have been elaborated by the Greek physiologists, though it is hinted at in the Hippocratic treatise, *The Sacred Disease*.

This method is employed by Malpighi in his experimental study of the tracheal system of the silk-worm. The experiments themselves are ingenious and, on the whole, satisfactory, but Malpighi's exposition of them (*De Bombycibus*, 1669) is not well ordered.

He seems to have had in mind, before starting his experiments, two suggestive facts and, also, an hypothesis drawn from a popular belief. The facts were, firstly, that insects like other animals die when deprived of air; and, secondly, that his silk-worm larvae, when immersed in hot water, gave off air bubbles, some arising from the general surface of the larva, but most coming from the direction of the stigmata (the openings of the tracheal system). This emission of bubbles seemed to him, moreover, an indication of some feature of the vital economy of the insect, for he noted that dead larvae gave off, in these circumstances, no air. The popular belief was that oil is harmful to insects.

The experiments were, therefore, designed to test two hypotheses at once: that the tracheal system is a respiratory system, and that oil is harmful to insects. The result is, as I have said, somewhat disorderly, but the following conclusions emerge with reasonable clarity:

(i) Oil is harmful to the insect only when it enters the stigmata. He writes: "I several times moistened with oil the belly, the head, the mouth and the back, leaving the openings of the trachea* free, and these I never saw die, nor even saw arise in them any signs of severe disorder."

(ii) The disturbing effects that follow the entry of oil into the stigmata are not due to any special property of oil as such, for butter and lard are equally effective. Nor are the effects limited to oily and fatty substances, for they are produced also by honey.

(iii) Long immersion in water is less harmful to the insect than is a brief contact with oil, if this enters the stigmata. From this fact Malpighi infers that liquids are harmful in pro-

* He consistently writes this word in the singular, meaning by it the whole tracheal system.

portion to the difficulty with which they are removed from the tracheal system.

(iv) Paralysis following the occlusion of the stigmata may be localized, in the sense that if only the front half of the larva is anointed (*viz.* from its head to the middle of its length) paralysis occurs only in that half; and, similarly, if the hinder half is anointed paralysis is limited to the hinder half. The value of these results would appear to be that they suggest an effect mediated by the tracheal system, the anatomy of which Malpighi had studied in considerable detail; they are consistent both with the hypothesis that the oil acts by preventing the entry of air, and that it acts as a poison. The decision between these hypotheses is given by the experiments of (i), (ii), and (iii) above.

Malpighi completes this experimental study by observations designed to show that the results achieved by the study of one species of insect could be generalized to other insects. He writes: "It seemed as well to try the same with insects of like kind, as cicadas,* crickets and so forth, in which death was again the outcome." This is not an instance of the use of enumerative induction, but a precaution against the possibility that the species used as an object of experimental study might turn out to be an anomalous case.

Malpighi draws the conclusion "It seems likely therefore that air continually enters and leaves this vessel (*viz.* the trachea) in silk-worms, as it does in such other animals as are possessed of lungs". The curious way in which he expresses this conclusion is due, I think, to his essentially physiological view of organs: he wants to assert that tracheae *are* lungs. Having given this result, he immediately continues, "Whether, therefore, a movement of the abdomen be necessary to the coming and going of the air, it seemed proper to enquire". He found that, in some insects, such

* Malpighi writes *in locustis*, but I think there is reason to suppose he is not referring to locusts.

movements of the abdomen were clearly visible. Moreover, he found in the cicada an apparatus of muscles and hard skeletal parts which appeared adapted to the production of these movements of expansion and contraction of the abdomen.

The results of this study were not in harmony with Malpighi's theory of the function of the lungs; for the structure of the tracheal system, as he himself correctly describes it, seems to exclude any possibility of accounting for respiration as a means of mixing "white" and "red" blood, even could fluids of these kinds be found in insects. In an obscure phrase, in his description of these experiments, he seems to say that he is at a loss to explain the purpose of respiratory organs. Mayow, however, in his *De Motu musculari*, 1674, drew from Malpighi's experiments on the silk-worm the conclusion that the essential element in all respiration is the introduction of air into the living system; and it is clear that Malpighi's results were an important factor in the formation of Mayow's theory of respiration.

It is clear that Malpighi's experiments on the silk-worm were in every way more complex than his experiments on the gall-bladder. In experimenting on this organ he had merely to discover whether it was or was not a necessary condition for the formation of bile, thus only one factor of the situation had to be explicitly considered; but in experimenting on the silk-worm he was concerned with at least two factors in each of two sets of experiments. In the first set of experiments he had to decide the question, "In connection with what part of the body is oil a cause of paralysis?"; and, in the second set, the question, "What property of the oil is responsible for the paralysis?"

The Greek physiologists were familiar with experiments involving only one factor, and the logical structure of such experiments had been fully elucidated by the Stoic logicians, but the use and understanding of experiments involving more than one factor seem to have developed after the time of Galen, who may be considered the last Greek physiologist. There is no reason to suppose that Malpighi originated the more complex type of experiment, but his use of it can, I think, be cited as an example of one way in which he was in advance of the Greeks in method.

We may consider here another way in which Malpighi's method appears to be novel when compared with that of the Greek physiologists; though, indeed, in this case his argument is based on observation, not on experiment. In his treatise *De Externo tactus organo*, 1664, he describes "pyramidal papillae" in the skin. These he compares with similar papillae which he had observed in the tongue, together with the nerve-fibres supplying them. He says that the papillae of the skin are probably organs of touch, and as evidence for this hypothesis he adduces the fact that they are the more numerous in those parts of the skin which are the more sensitive to touch.

Since this sort of inductive argument is a refinement of common-sense arguments used by all human beings; it cannot be said that it is never employed by Greek physiologists; but Galen, whose writings are the only important source for our estimation of Greek physiology, seems peculiarly insensitive to the *precise* use of all forms of quantitative argument. It is certainly the case that, when we try to discover exactly what is new in Harvey's arguments for the circulation of the blood, we are particularly struck by his measuring the capacity of the heart and estimating the amount of blood emitted during precise periods of time.

From Malpighi's description of the corpuscles or "papillae" of the skin, it seems that he is referring to what are now called "Meissner's corpuscles", and it is interesting to note that Malpighi's argument as to their function still has some force :

"Situated quite superficially in the dermis, and often within the dermal papillae, are 'Meissner's corpuscles'. Since they are mainly found in cutaneous areas which are particularly sensitive to touch, they have been regarded as 'tactile corpuscles'." (W. E. Le Gros Clark, *The Tissues of the Body*, Oxford, 5th edn., 1965, p. 323.)

MALPIGHI'S WORK ON PLANTS

I have said above that Malpighi wrote the first book on the micro-anatomy of plants. This statement requires qualification, for his *Anatome plantarum* of 1675 is not entirely, or even mainly, concerned with what can be called micro-anatomy if the expression is used in a precise sense; moreover, the chronological relation of Malpighi's studies on plants to those of Grew, whose book *The Anatomy of Plants* appeared in 1682, is complex.

With regard to the first point, it is necessary to observe that a great deal of Malpighi's work on plant anatomy is concerned with structures which, though small, can be seen with a simple lens of low power, or even with the naked eye. However, the expression "micro-anatomy" is not altogether inappropriate to the whole work, because, as Sachs acutely remarks, many things which could have been seen with the naked eye were not in fact seen until the use of the microscope had accustomed scientists to give their attention to the smaller parts of organisms.

The temporal relation and the question of precedence between Malpighi and Grew have been thus analysed by Sachs :

"About the time of the appearance of Hooke's *Micrographia*, 1667, Malpighi and Grew had already made the structure of the plant the subject of detailed and systematic investigations, the result of which they laid before the Royal Society in London almost at the same time in 1671. The question to which of them the priority belongs has been repeatedly discussed, though the facts to be considered are undoubted. The first part of Malpighi's work, the *Anatomes plantarum idea* . . . is dated Bologna, November 1, 1671; and Grew, who from 1677 was Secretary to the Royal Society, informs us in the preface to his anatomical work of 1682, that Malpighi laid his work before the Society on December 7, 1671, the same day on which Grew presented his treatise *The Anatomy of plants begun*, in print, having already tendered it in manuscript on the eleventh of May in the same year. But it must be observed that these are not the dates of the larger works of the two men, but only of the preliminary communications, in which they gave a brief summary of the researches they had then made; the fuller and more complete treatises appearing afterwards; the preliminary communications formed the first part of the later works and to some extent the introduction to them. Malpighi's longer account was laid before the Society in 1674, while Grew produced a series of essays on different parts of vegetable anatomy between 1672 and 1682; and these appeared together with his first communication in a large folio volume under the title *The Anatomy of Plants*, in 1682." (Sachs, *op. cit.*, pp. 231–2.)

The history of these publications is even more complicated than Sachs allows, for Grew's first publication, actually called *The Anatomy of Vegetables Begun*, printed by an order of the Royal Society dated 9 November 1671, bears the date 1672, though Grew asserts that "The Book being quickly printed off; I ordered it to be Presented to the *Royal Society*; which was accordingly done at one of their Meetings *December* 7, 1671". However, there need be no real discrepancy in this, for if the book was not printed till December 1671, the printer might well put upon it the date 1672 as the date when the book would be offered for sale.

It is not possible to assess here the relative merits of Grew and Malpighi. On one important question Grew is in advance of Malpighi, for Malpighi regarded the pollen as a mere waste-product, and the stamens as organs for excreting this material and so refining the food destined for the embryo-plant in the seed (*Opera omnia*, Lugd. Bat., 1687,* p. 70). The notion of local refinement of food destined for particular organs is clearly adopted from Galenic physiology. Galen asserts that four faculties, which he calls the "handmaids of nutrition" are necessary "for every part which is to be nourished". These handmaids are the faculties of "attraction", "rejection", "alteration", and "retention" (*On the Natural Faculties*, III, ix). These four faculties serve the final purpose of "assimilation", and Galen rightly points out that each part of the body must perform assimilation for itself. The whole set of ideas is clearly not without value, but Galen thought of rejection and expulsion of the waste products of assimilation as too closely analogous to the production of faeces, and consequently

* The date is incorrectly given by the printer: the book must have been printed after 1698.

expected that the quantity of rejected matter would commonly be large in proportion to the quantity assimilated, and as a result of this expectation he too easily persuaded himself that the body is provided with several large organs whose sole function is elimination. Thus he supposed both the gall-bladder and the spleen to be concerned with the elimination of the waste products of the process of blood formation in the liver. Within the context of Galenic physiology it was quite natural to suppose that food destined for the embryo must be especially "refined" and that the waste products of the refining process would be considerable in quantity and locally eliminated. In the case of the flower, preparing food for the embryo in the seed, the stamens seemed the obvious organs of elimination. On the "use" of the stamens Grew adopted a curious compromise. He accepted Malpighi's suggestion that they are organs of elimination, even accepting Malpighi's analogy of the menstrual discharge; but he considered that the pollen is "the male seed" or sperm (*The Anatomy of Plants,* IV, v).

The first part of Malpighi's *Anatome plantarum*, 1675, was followed some years later (1679), by a second part containing three extremely valuable sections on the germination of plants, on galls, on plants which grow on other plants.

In his study of germination he presents excellent drawings of the stages of the development of the seedling. Seeds and seedlings of the gourd (*Cucurbita*), of beans and peas, of corn and millet are excellently represented, even the tubercles on the roots of the leguminous plants are clearly recognizable in his drawings.

He noted that in some plants, as, for example, the gourd, the cotyledons develop fully into leaves, whereas in others they retain their initial forms. He asserts that the cotyledons "play the part of a placenta" in supplying nourishment to the young plant. He confirmed this by showing that after removal of the cotyledons the seedlings grow feebly and soon die. He also showed that the effect is quantitative, for if only one of the two cotyledons be removed the seedlings grow more slowly than with two, but do not perish, as those do from which both cotyledons have been removed.

He tried the effect of watering seedlings with salt solution, wine, human urine, and acids and alkalis, and showed that they suffered more or less according to the fluid used. He also showed that seeds do not germinate under water covered with oil, or reach only the stage of the emergence of the root-tip. Möbius was puzzled by this experiment, but it seems clear that Malpighi wished to expose the seeds to water in the absence of air, since already in 1671 he had decided that air was necessary to the life of plants.

The treatise on galls is perhaps the most nearly perfect of all Malpighi's works. He mentions no less than sixty-seven galls, of which most are minutely described and carefully drawn. In very many cases he is able to represent within the galls the larvae of the insects which cause them. Malpighi's descriptions and figures are so excellent that in almost all cases it has been possible to identify the type of gall and the particular species of insect now known to cause it. Malpighi says that the various galls are caused by "various kinds of fly". He gives a figure of one of these, and his figure is so clear and detailed that it is possible to recognize the insect as belonging to the genus *Rhodites* (this figure is reproduced in Singer's *History of Biology*, 1950, p. 155). He represents the ovipositor most carefully, and is fully aware of its nature and function. He describes how he saw a gall-wasp in the act of oviposition, how he found the egg and compared it with those still in the oviducts of the insect.

In his treatise on plants which grow on other plants, Malpighi carefully describes and figures mistletoe, lichen, liverwort, moss, and several types of small fungi. He says it is probable that lichens and fungi are spread by the wind and that they are found principally on the northward side of trees, because "they take root only with difficulty on the opposite side, since the fluid that nourishes them is evaporated by the rays of the sun".

It is impossible to give more details concerning Malpighi's studies on plants. Rather than attempt a lifeless catalogue of his observations, I offer the following translation of the last few paragraphs of his preliminary essay of 1671, which illustrates both the extent of his knowledge and the acuteness of his interpretations.

"As an endpiece it would be as well to add something concerning the economy of plants, the use of their parts and, particularly, on the path taken by their food. But, since the undertaking is full of hazards, because of the uncertainty and near impossibility of the observations *required*, I will only make some few suggestions, that those who are more studious and more practised *than I am* may propose something better.

"Firstly, I may take it as established with certainty, from what has been said, that there are in plants many vessels or tubes; for it is clear from what has been noted above that there are in the bark (*cortex*) a great many fibres hollowed out in the manner of vessels. Similarly, we have found the wood to be made up of fibres, of which we have declared some to be composed of a spiral band, while others are built up of, as it were, discs open *at their ends*; and we found, lastly, also a special *kind of* vessel from which milk or resin flows.

"But at once the doubt arises, whether the fibres are of the same nature and made for the same purpose in both bark and wood. The examination of herbs, indeed, and of plants such as giant fennel shows it to be probable that they are the same, since the fibres of the bark, having become hardened, change into wood. It is certain that both kinds of vessel abound in fluid, as appears from its flowing out of *them*. I can also show that much air is certainly contained in plants, for when cut under water they yield a great deal of air, and there are within them tubes extremely like the lungs of insects. It is doubtful whether air is drawn in through the lower and last roots, or through the upper ends *of the plant* which are in contact with the air. Since, however, the spiral tubes or tracheae play a great part in the composition of the root, and air is the more readily expelled in an upward direction, and its operation and efficacy tend upwards, so I should judge it the more probable that air and exhalations are led from the lower parts of the soil.

"I can also show that a juice is clearly to be seen, sometimes of the appearance of milk, sometimes of gum, sometimes of terebinth, having a vessel of its own; and I thence infer that each plant has a juice peculiar to itself and carried in its own proper vessel.

"Concerning the motion of the said humors, however, and the places towards which they flow, I consider it certain that they can in many *cases* be inverted, since a shoot of the fig, the plum, the bramble, etc., if struck into the earth may put forth roots at the thinner end and may grow into a tree, though less tall *than usual*; hence the pathway of the food is *in such cases* inverted.

"I should like to propose that there is just as evidently a mutual anastomosis between the vessels of alimentation, so that there is also a sideways passage of the juice, which consequently does not travel always in the same tube. This is obvious when some of the little ribs of the leaves of the gourd, the lemon, etc., have been broken, and the substance beyond, in line with the broken tubes, continues to grow and live, food being provided by tubes at the sides. This proposition, *that there is lateral communication between the tubes*, is corroborated by the reticular plexus of vessels *to be found* in leaves, bulbs, bark and wood.

"It will also appear that, both in the tubes of the wood and in those of the bark, fluid goes up towards the branches and leaves. It will be clear, however, that the ascent of the fluid will not occur in each and every one of the tubes, if we consider the use and activity of those leaves of the gourd which are formed out of the flesh of the seed, which some call 'the pulp'. For it is certain (as will appear from the day-to-day record of the germination of beans) that the two leaves *formed* by the spreading out of the flesh of the seed, and to be found also in gourds and some other *plants*, receive fluid from the root, since they increase enormously in size; but it is clear also that they send *fluid* back into the stem for germination and growth. For if they are broken off from the germinating plant, it shrivels; and in beans the flesh of the seed, which is a leaf squeezed up into a ball, although it does not spread out and become green, yet, having received a watery fluid *from the root*, it pours out an oily substance, the proper juice of the little plant to whose stem it is attached, so that it acts the part of a placenta.

"Hence I infer it to be probable that leaves have been formed by Nature for the following purpose: that in their cells (*in ipsorum utriculis*) the contained nutritive juice brought to *them* by the woody fibres may be elaborated; for the fluid, well mixed during the long journey through the numerous anastomoses of the vessels, and refined by the power of the sun's rays, is combined with the material already long contained in the cells, and suffers a new combination of parts and an evaporation *of its fluids*; very much as happens to the new [*viz.* just eaten] food of animals, which, infused into the other blood left in the vessels by an *earlier* act of nutrition, is *itself* raised by it to the nature of blood. I can also show it to be probable that the same state of affairs is to be found in the cells of the bark, and in the sets *of cells* running transversely in the wood; so that it does not seem unreasonable to infer thence that the nourishment of the plant is concocted and conserved in these little parts *also*.

"The juice of the pericarp, collected and elaborated in cells of the cortex, and the little sacs of onions containing a deal of fluid, which they expend upon the growth of the new germ, throw much light on what has already been said; as do also the swollen cells in the severed stump of an oak, and of other *trees*, which are so stuffed with the stored fluid, which would normally be consumed as food for the trunk and branches, that the region of the woody fibres is considerably augmented, and the tubes themselves are contorted.

"The pathway taken by the elaborated food, by which it flows out from the cells of the leaves, of the bark and of the pith, and to some extent flows back *to them*, must remain conjectural. Though it is very likely that it flows down into the trunk and roots. However, the terminal parts of the branches beyond the leaves, which *parts* possess cells and the like, do not appear to be deprived of this benefit. Hence I offer it as probable that the already elaborated nourishment flows out of the cells through the reticular vessels, of which indications are observed in reed-mace, without *any* constant or considerable flux and reflux, and that it remains stagnant, as it were, in the lactiferous or sanguineous vessels, and flows thence according as it is needed, the need being provoked by transpiration and by various external influences; sometimes flowing to nearby and somewhat higher parts; and thus supervene growth and nutrition, while, also, many branches of trees and many herbs lie horizontally, *and hence are even more easily supplied than are the vertical shoots.* Otherwise a simpler path may find favour; for I long believed the matter of the nourishment to be carried through the woody fibres from the roots to the whole mass of the tree, and that, from the transversely arranged cells, a fermentative juice was mingled with *the nourishment*; such a juice as is poured from the glands *of animals* into the passing chyle and blood : but further thought showed me that the way suggested above is the more probable, *that is, from the cells of the leaves to the other parts of the plant.*

"Having given a brief sketch of the foregoing matters, I will suggest the possible uses of air in plants, having thought upon the tracheae of insects, which carry air to all parts of the body : it is extremely probable that, over and above the communication of life, it is by means of air that plants are enabled to grow upwards during germination and are held upright during *the rest of* growth and the *whole* course of *their* life; and by air that the ascent and fermentation of juices are facilitated, with other *operations* of the same kind.

"These, then, are *the thoughts* which *my* mind, *my* body being sick, meditates for its consolation, and which I present to you without much order, which, if they be not altogether worthless nor yet too far from the straight path of philosophizing, shall be evaluated by your sound judgement; I will expose them more fully to observations repeated again and again, and will present them to your minds as an offering, though an unworthy one."

Several features of this passage are of great interest. It seems legitimate to translate *utriculi by* "cells", since Malpighi was in fact looking at what we now call "cells", and he was clearly in no doubt as to the fact of their being the active elements of the plant. His supposition that the food of the plant is elaborated in the cells of the leaves, and the argument drawn in support of this supposition from the cotyledons of seeds are of great merit, and better than most of the speculations of the next hundred years. The fundamental physiological ideas with which he is operating display an interesting transition between Galenic and more recent physiology. The idea of food being changed into blood is essentially Galenic, but the emphasis on the activity of glands is hardly to be found before the seventeenth century. The reference to the sun as an active agent in plant metabolism is, I think, exactly transitional, for in Greek physiology the active agent in such processes was "warmth", but the "warmth" was usually considered to be "innate", not derived from outside the organism. Galen's theory of the maintenance by breathing of the "warmth" in the heart is, however, though muddled, already greatly in advance of Aristotle's belief in a purely "innate warmth".

MALPIGHI AS EMBRYOLOGIST

We have seen that Malpighi studied the development of plants, and it is now necessary to say something about his studies of animal embryos. He wrote two tracts on the development of the chick. Both are dated 1672; indeed, the second is called an appendix to the first, though it might well stand as a separate work. The first, dated February 1672, is entitled *De Formatione pulli in ovo,* the second, dated October 1672, has as its shorter title *Appendix de ovo incubato.*

As with the work on plants, it is here also very difficult to give a brief account of what is essentially a record of innumerable observations. Malpighi's account of the development of the chick illustrates the general statement that the microscope of the seventeenth century, primitive as it was, opened up a new world to observation. This is particularly clear when we compare Malpighi's figures with those of Fabricius's *De Formatione ovi et pulli,* 1621. Fabricius says that by the fourth or fifth day of incubation all the essential parts are visible except the limbs; but it is not till about the eighth day that he is able to present figures of the embryo, and he can see little beyond the mere outline. He gives one figure which belongs, perhaps, to the fifth day, but this figure is so rudimentary that it might be copied by a single stroke of a reed-pen. It has the merit of suggesting a fish-like outline. Malpighi's figures, particularly those of the *Appendix*, represent in some detail the condition of the embryo at the end of the first day, or slightly earlier. He has two clear and clearly recognizable figures showing the embryo with respectively seven and eight pairs of somites. At three days he shows the optic vesicle, three ovoid cerebral vesicles and the medulla and spinal cord, and the rudiment of the heart. He follows well the development of the brain and heart to the tenth day. On the third day he sees three branchial arteries with their terminations in the heart and aorta.

The style of Malpighi's text is strikingly different from that of Fabricius. Fabricius is discursive and philosophical and frequently appeals to authorities, whereas Malpighi hardly deviates at all from simple description of his own observations. I can find no evidence that he was greatly interested in general theorizing about the nature of generation, and to classify him without qualification as a "preformationist" seems to me a considerable anachronism; for the sharp distinction between "preformationists" and "epigeneticists" cannot be said to have existed before the eighteenth century, and he is far from indulging in the absurdities of the dogmatic preformationists. To understand the position of Malpighi and of his contemporary Swammerdam it is necessary to start, not from ideas current in the eighteenth century, but from the speculations of Fabricius and of Harvey. Fabricius, using arguments later reproduced in a modified form by Bonnet in the middle of the eighteenth century, argues that the formation of one part necessitates the formation of another, and that this dependence may be mutual: he argues, for example, that the liver cannot exist without the heart, nor the heart without the liver. For various reasons, some of which are highly fanciful, he concludes that all the major internal organs must be present from the start. But he does not mean that they must be present and also fully formed from the first. Fabricius's opinion seems to have been adopted by Malpighi, but it is not, as has been suggested, an opinion which conflicts with his observations. For it does not exclude development and change of shape of the organs. Moreover, since its most rational basis is an argument which is in substance essentially physiological, it concerns principally organs which have begun to function : thus it does not exclude the possibility of almost shapeless rudiments being present before the formation of functioning and visibly distinguishable organs. The theory of Fabricius, then, when reduced to its essentials, is only "preformationist" in the sense that it requires rudiments of all the principle organs to be present from the first. The notion has the considerable merit that it supposes the fertilized egg to be already organized, and to possess an organization which is causally connected with the subsequent formation of the embryo.*

I do not mean, of course, that an analysis of the kind just presented was consciously made by anyone in the seventeenth century; but I think it certain that an obscure awareness of the consistently causal nature of the Fabrician theory accounts for its being preferred to Harvey's negativistic attitude towards a scientific embryology. Harvey, in his *Exercitationes de generatione animalium*, 1651, says of the developing chick that "it requires and makes a material which is ... various in its nature, and variously distributed, and such as is now adapted to the formation of one part, now of another; on which account we believe the perfect hen's egg to be constituted of various parts" (*Exercitatio*, 45). This is excellent as far as it goes, but Harvey is anxious to exclude naturalistic explanations of development. He says : "It is a common mistake with those who pursue philosophical studies in these times, to seek for the cause of diversity of parts in diversity of the matter whence they arise" (*Ex.* 11). He speaks of "inherent mind, foresight and understanding, which from the very commencement to the being and perfect formation of the chick, dispose and order and take up

* Fabricius did not believe that the embryo was preformed either in the unfertilized egg or in the sperm. He believed it to be *generated* at the time of conception. His ideas only look like preformationism because he seems to have thought that all the principal organs were generated at once at the moment of conception.

all things requisite" (*Ex.* 57); and he says that the formation of the embryo exceeds the powers "not only of the elements, but of celestial elements* and even the soul, vegetable animal or rational" (*Ex.* 71). Thus Harvey despairs of any but a purely descriptive embryology, and it was natural for the scientists of the late seventeenth century to turn back to the ideas of Fabricius which, however quaint their form and expression, contained a nucleus of naturalism and the hope of an ultimate causal study of development.

This, in brief, is the general theoretical background against which we have to view the little that Milpighi has to say about the fundamental nature of development. To understand the more particular terms of the theorizing of this period we have again to turn to Harvey.

Harvey seems to have coined the word "epigenesis" and he contrasts the idea signified by this word with that signified by the well-established word "metamorphosis". He says (*Ex.* 45) that there are two ways in which something can be made out of something else *tanquam ex materia* : these are "metamorphosis" and "epigenesis". In metamorphosis the whole mass of matter is given together and is shaped into the thing made; as in the making of a statue from marble, or the production of an adult insect from its larva. In epigenesis the material is made and shaped at the same time (*materia simul formatur et fit*), as one who works with clay makes the same shape as the sculptor working in stone, but does so by adding piece to piece and shaping the pieces as he adds them. This is the mode of production of the more perfect animals. Clearly, the analogy of the artist working in clay is incomplete, for he does not *make* the matter as he forms it. Harvey means that special kinds of matter are elaborated in the higher animals, starting from the less elaborated matter of the *primordium* or of the food supplied by the maternal blood. Harvey says :

"In the first way occurs the generation of insects, where a worm

* This is a reference to Aristotle's *pneuma*, which was "allied to the element of the stars".

is born by metamorphosis from an egg, or from putrid matter primordia are spawned . . . from which, as from a caterpillar which has reached its full growth, or from a chrysalis, a butterfly or a fly arises of the exact size (*justa magnitudine*), and no bigger than at its first beginning.

"But an animal which is created by epigenesis attracts, prepares, elaborates, and makes use of the material all at the same time; the processes of formation and growth are simultaneous . . . the formative faculty of the chick rather acquires and prepares its own material for itself than only finds it when prepared."

It was not against the notion of epigenesis, but against that of metamorphosis that Swammerdam and Malpighi protested. Swammerdam's book, called in its Latin form *Historia generalis insectorum*, was published in Dutch in 1669, in French in 1682, and in Latin in 1685. These dates are important, for it is unlikely that Malpighi's work published in 1672 could have been influenced by Swammerdam's.

Swammerdam is scandalized by Harvey's notion of metamorphosis, partly because Harvey combines it with a reference to spontaneous generation, which, Swammerdam asserts, "does violence to the order of Nature, which is most constant"; for he thinks of spontaneous generation as generation by chance. "It seems to me," he says, "the grossest abuse to make subject to chance that adamant and immutable constancy of Nature." He asserts that insects arise only from eggs laid by other insects of the same species.

However, nothing could be further from the truth than to represent Swammerdam's book as a theoretical treatise. He is only forced to appeal to general principles by his horror at Harvey's mistakes. Indeed, he explicitly issues a warning against trusting to reason : "We should take all possible pains to discover the truth by observations made upon natural objects, rather than to seek it in our reason, which is naturally weak and subject to error." Among the truths which he discovers by observation is that the form of the butterfly may lie hidden within the chrysalis, and that rudiments of the organs of the butterfly may be found even in the caterpillar. He is much impressed by this discovery and tends to generalize his results, stating, moreover, that "the single principle of all the changes which occur both in the eggs of insects and in the worms, or the caterpillars into which they develop, depends upon a clear and distinct knowledge of the nymph". But he says that the limbs "grow in the worm little by little just as they do in other creatures" and that "they grow very slowly, one after another". Moreover, he frequently refers to the growth of limbs as "epigenesis". Indeed, he is so far from rejecting epigenesis that, in effect, he blames Harvey for not realizing that it occurs in insects just as much as in the higher animals. What he says about the initial formation of the egg in the mother is difficult to interpret : at one point he says that it depends upon "an invisible principle, although one which really exists"; but, in discussing whether injuries suffered by the parent could be inherited by the offspring he says : "Since the offspring is already preformed when the parent is born, no accident that can occur to the parent can affect the nature of the offspring", he adds that this will help us to understand how the whole human race is stained with original sin, and how Levi could be "in the loins of Abraham" (Hebrews vii. 10). He admits, however, that this last reflection was suggested to him by Malebranche. It seems not improbable that the notion of the offspring being preformed when the parent is born was also derived from Malebranche, for it is more speculative than the strongly empirical bias of Swammerdam would lead us to expect from him, and is certainly not clearly present in the other passages in which he refers to the formation of the egg. What is certain, however, is that Swammerdam did not think that preformation meant the preformation of the perfected adult form, so that his preformation is something like an extension of the "pre-formationism" of Fabricius, and does not exclude the development of parts. As I have said above, it is most improbable that Malpighi, when he wrote his papers of 1672, knew of Swammerdam's observations or of Malebranche's speculation. Even if he accepted the kind of preformation favoured by Swammerdam, there is no justification for Cole's assertion that he "could see one thing and believe another" (F. J. Cole, *Early Theories of Sexual Generation*, 1930, p. 48). Nor does the sentence quoted by Cole express a belief in a preformation so radical as to exclude that development of organs actually observed by Malpighi. "*Quare pulli stamina in ovo praeexistere, altioremque originem nacta esse fateri convenit, haud dispari ritu ac Plantarum ovis*", states no more than that the rudiments of the chick pre-exist in the egg as those of the plant do in the seed. This assumes a "preformation" no stronger than that proposed by Fabricius. Moreover, if we glance at what Malpighi had already written about "the eggs of plants" we find that he believed the see to contain no more than the rudiments of root stem, and leaf.

"So the seeds of plants are not analogous to the eggs of oviparous *animals,* since seeds require after *their* incubation food, which is entirely included in the egg, *but which they have to derive from the soil.* They do not seem to differ greatly from the eggs of viviparous *animals,* since in both the spine together with the structure of the living thing pre-exists concisely, and almost the whole of the material for growth comes from without. They seem to differ only in this, that the structure of the little plant in the seed, which has been described above, is more obvious than the rudiments (*inchoamenta*) which occupy the cicatriculum* of animals; and in the womb of the young egg of viviparous *animals* undergoes continuous growth and change (*mutationem*), juices from the womb having seeped into it continuously. However, the seed differs less from the others, *that is* the imperfect eggs, which suffer a swelling of their contained fluid, particles having entered *them* from without; now this, a ferment having been set free by the humid particles, makes the egg, that constitutes the little plant, swell up so that it quickly breaks out of the husk.

"From this it is allowable to conclude with probability that the seed of plants is an egg containing an embryo made up of the more essential parts (*foetum principalioribus compaginatum partibus*), and which can, moreover, be kept alive for years until,

* The "germinal spot" visible on the yolk of the hen's egg.

the external moisture having entered and swelled it, the parts are more fully unfolded, and so the little plant is said to sprout. The details of this will be made clearer by the daily study of sprouting seeds, of which, God willing, we will give the description and drawings, with the drawings already promised of the other parts."

This is the concluding passage of the *Anatome plantarum*. I pass over several difficulties in the Latin, which affect the meaning only trivially.

The passage contains some interesting vestiges of Greek biology. Thus the "imperfect eggs" (literally "eggs of *imperfect animals*" but this seems to be a *lapsus*) are those distinguished by Aristotle from the "perfect" eggs of birds and lizards : the "perfect" eggs do not increase in size during development; the "imperfect" eggs, principally those of marine animals, do increase in size. The observation is, of course, correct, only the terms are peculiarly Aristotelian. The "spine" (*carina*), which Malpighi supposes to pre-exist as a rudiment in the egg, recalls a Galenic notion that Nature first lays down the spinal column and ribs as a shipwright first lays down the keel (*carina*).

If we now restore to its context the sentence, "*Quare pulli stamina*, etc", we shall have complete the greater part of what Malpighi says about "preformation", and shall be able to appreciate how far he is from the more absurd forms of the doctrine : all he has done is to extrapolate from what he has observed in the seeds of plants, putting back into the egg at its first inception a smaller and, we may infer, a still simpler version of the rudiment observable in the seed. This, though mistaken, is neither absurd in itself nor inconsistent with his actual observations.

The first of Malpighi's two papers on the development of the chick begins thus :

"In the making of machines, the various parts are wont to be formed by a previous labour, so that those which are later to be joined together are seen separately. Many of those who are adepts *in the ways* of Nature and solicitous of the investigation of animals had hoped that in her works she would proceed in the same manner; for, since it is most difficult to analyse the intricate structure of the body, it would be of great assistance to inspect in their separate primordia the various parts produced. But I fear that the *course of* mortal life lies between uncertain limits and its starting point is as obscure as its goal. For just as death, so Tully instructs us, belongs neither to the living nor to the dead, so, I think, something of the same kind occurs in the first beginning of animals; for while we are anxiously seeking the production of animals from the egg, we find in the egg itself something like an animal (*ferè animal*) already established, so that our labour is in vain; for, not having found the first beginning, we are forced to await the appearance of the parts as they come forth in turn.*

"In this investigation, indeed, many have laboured (*insudarunt*), among whom stands out your immortal Harvey whose perfected observations still so far instruct the world that they brush aside my labours, particularly mine, as superfluous.

"Since, however, as he has said himself, 'the first rudiments (*stamina*) of nature lie, for the most part, veiled in deepest night, and by their subtlety they evade the acuity of the mind no less than that of the eyes', and the so greatly varying power of Nature, as though the time of ripening *were* uncertain, now hastens, now retards the egress of the foetus; so, most learned Fellows†, suffer me to communicate to you certain rough beginnings of observations from the study of incubated eggs, the which I purpose often to repeat. . . . Among the parts of which eggs are constituted, the cicatricula, or round spot, takes the first place; for, so it seems, it is for the sake of this *round spot* that the other parts have been produced *and associated* with it. Its wonderful structure, therefore, offers itself for investigation, and I will describe briefly its principal changes and its *various appearances*. . . . In eggs laid the day before and not yet incubated (as I saw last August, the heat being

* Reading *emergentium* for *emergentem*, in "emergentem successive partium manifestationem expectare cogimur".

† Fellows of the Royal Society of London.

very great) the cicatricula has the size of the figure A,* here roughly sketched by me, in the centre being a little sac. . . . Then in the little sac as in an amnion, when I exposed it to the sun's rays, I perceived the enclosed foetus, of which the head, with the appended rudiments of the spine (*cum appensae carinae staminibus*) was clearly visible. . . . Hence the rudiments of the chick pre-exist in the egg, and we must allow them *to be* derived from some earlier origin, just as in the case with plants."

Malpighi's drawing of "the enclosed foetus" shows an object like the head of a match broken off with a very short piece of the wood attached to it. There is no question of his having believed himself to see a miniature chick; nor did he think such a thing to exist at this stage, as is proved by the phrase concerning the emergence of the parts *successivè*.

A final reference to generation occurs in a work written by Malpighi somewhere about 1690. Answering a critic of his ideas he writes : "I await with composure of mind for him to instruct me and at the same time to prove that metamorphosis *occurs* in some animals, and to bring proofs against the pre-existence of the principal parts in the egg." We note that it is still "the principal parts" which are said to pre-exist, and that preformation is contrasted with metamorphosis, not with epigenesis.

CONCLUSION

The position of Malpighi is not difficult to assess from a consideration of his works, but we have the additional assistance of the long paper written by him some three years before his death and published posthumously in London in 1697.

This paper is a reply to an anonymous critic and displays signs of considerable irritation. Nevertheless, Malpighi sums up his own position with great clarity. He points out that, as an anatomist, he is in no way departing from the spirit of Galen, whose work he is continuing; and he proves by numerous citations that Galen's method was the same as his own. He shows

* The figure shows it to have been about 3 by 6 mm.

himself a moderate preformationist; and he explains in detail and with great lucidity how he thinks physiological work should be conducted. He sees no rivalry between physical and chemical explanations, but considers them to be complementary. He is a Cartesian, likening the relation of soul and body to the relation of a living being to a mill that it sets in motion. The physiologist is free to give a mechanistic explanation of the body, just as of the mill, without being obliged to explain the being that uses the mill or who uses the body.

He advocates the employment of models as explanatory devices, and he here uses the word *modelli* (he is writing in Italian); and gives several examples of successful models, including the *camera ottica* used by Kepler to explain the formation of the image on the retina of the eye.

Just as he avoids the errors of the extreme preformationists in embryology, so he avoids the uncritical attitude of those who, like Boerhaave, believed that any model which could explain the phenomena could be assumed to represent the underlying physiological mechanism, and that no further tests of the validity of the model need be applied. Having expressed himself as an enthusiastic adherent of what is in effect Cartesian physiology, he issues a warning, which is at the same time an apology for his devotion to observational anatomy :

"And since the progress of the human intellect, if it rests upon mere hypotheses (*fondata sulle sole hipotesi*) is uncertain, although it may be probable, as with the 'man' of Descartes, so it is safer to base it upon things perceptible to the senses."

(*Note.* This essay was written before the publication of Adelmann's superb editions and translations of the works of Malpighi. Hence the absence of acknowledgements to that most distinguished authority.)

CHAPTER 3

CHRISTOPHER WREN, 1632–1723

A. J. PACEY

ARCHITECTURE IN THE ROYAL SOCIETY

It is probably inevitable that Christopher Wren should be something of an enigma to modern students of the Royal Society. His activities in science and in architecture seem too unrelated to fit into the framework of learning which is familiar today, with its many specialized subjects, and it is tempting to confine attention to either Wren the scientist or Wren the architect according to the direction of one's own interests.

But although Wren's work seems exceptionally diverse by modern standards, it was not nearly so exceptional among the early members of the Royal Society. Hooke's career was in some ways a close parallel of Wren's, embracing both architecture and science, and involving frequent collaboration with Wren in both fields. Evelyn, a secretary of the Royal Society, was also interested in architecture but more as an amateur. Wren was exceptional, not so much in taking up architecture, but in becoming so successful as an architect. Thus the question of Wren's diversity applies to other Royal Society members, and we must ask how architecture would have fitted into the Royal Society's programme; whether as a technology or in some other way.

The first point to be made is that the science which the Society sponsored in its early days was in any case very diverse. Most Royal Society members had wide interests and a rather unsystematic approach, and there was no question of specializing in one self-contained subject. There was no sharp distinction between pure and applied sciences as we know it today, but there was an implicit distinction between applied sciences that had become accepted as part of the body of learning, and those which remained the province of the artisan. Thus while medicine and navigation were based on learning of an academic kind, mining, metallurgy, and textiles were largely left to the skills of craftsmen.

Now at first sight it seems surprising that the Royal Society was affected by such a distinction, because the Society's members believed strongly in the practical value of science and were anxious to apply it to any industrial trade in which it might be of use. This sense of the social function of science was clearly expressed by authors like Boyle and Sprat, and in the information about various trades which the Society collected. Such information was contained in the "Histories of Trades" compiled by members, and embraced the whole industrial activity of the nation—mining, iron-making, textiles, paper, brewing, and bread-making. These histories of trades should have yielded information as to where industrial development was being held up by technical backwardness, so that the expertise and inventiveness of Royal Society members could be focused on such problems. But all this enthusiasm for improving trades had disappointingly little effect.

Medicine and navigation were the applied sciences upon which the Royal Society made its greatest impact and they were also sciences in which Christopher Wren was active. In his inaugural lecture as Professor of Astronomy at Gresham College, he emphasized the usefulness of that subject, and spoke of the inheritance which had "fallen to Mankind by the Favour of Astronomy. It was Astronomy alone that of old undertook to guide the creeping Ships of the Ancients, whenever they would venture to leave the Land to find a neighbour Shore".

Wren's attitude as he expressed it here was fairly typical of the Royal Society at its foundation. His interests were no more technological than most other members of the Society, although, like them, he was interested in the practical applications of science in so far as this was possible within a context of learning. But Wren, like most other members, was not very interested in the industrial technologies.

Architecture, in Wren's circle, was certainly not a technology in the modern sense of the word, but neither had it the remoteness and obscurity often associated with a fine art. Like medicine and navigation, it was a subject whose practice depended on learning that was established and respected. Admittedly, it had only just attained this status, for it was only a generation earlier that the design of buildings had become accepted as a learned occupation. Admittedly, too, the scientific content of architecture was small, limited to structural problems and the geometry of surveying, drawing, and proportion. Where scholarship was most important was in the study of Roman and other classical buildings.

But the part architecture played in the material side of civilization, and its roots in a respected body of learning, were sufficient to bring it within the Royal Society's range of interests. After all, if the Society's members were supposed to cover "all the Works of *Nature*, or *Art*, which can come within their reach" it should cause no surprise that two of them spent much of their time designing buildings.

CHILDHOOD AND YOUTH

Wren does not appear to have shown much interest in architecture until after the age of 30, but it is probable that his father, the Rev. Christopher Wren, had an ability in that field. While Rector of East Knoyle in Wiltshire he supervised some minor architectural work in his church and he designed a building for Charles I intended as an addition to Windsor Castle. At some time, too, he had read and annotated Sir Henry Wotton's *Elements of Architecture.*

Wren was born at the East Knoyle Rectory in 1632. When he was 2, his father became Dean of Windsor and Registrar of the Order of the Garter. The young Christopher was at Westminister School during most of the Civil War, and so was partly shielded from the trouble which came to the family through their Royalist connections. His uncle, the Bishop of Ely, was imprisoned, and his father's house was pillaged in a search for official papers.

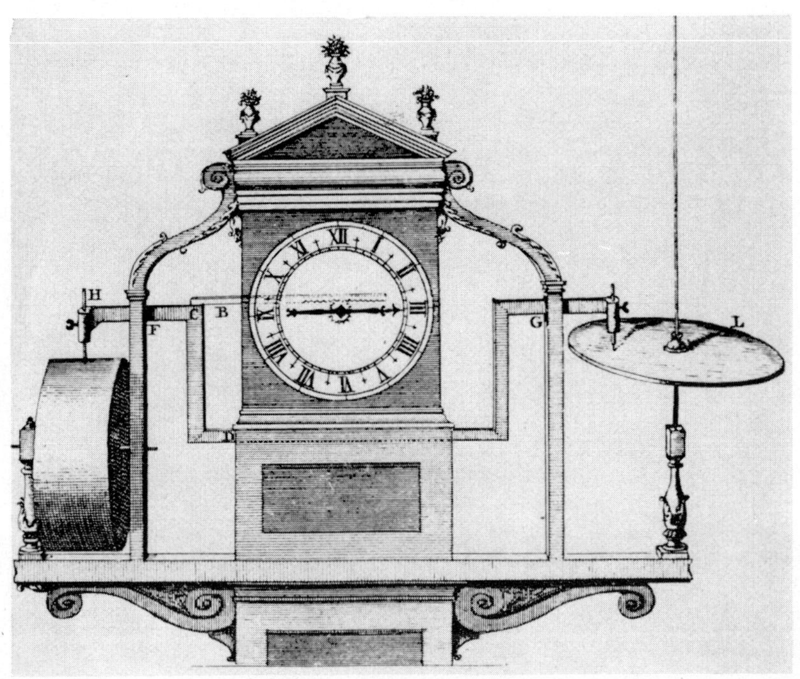

Fig. 3.1. Weather clock.

But the Registers of the Order of the Garter were successfully hidden together with the George and Garter decorations. The decorations, which included a large number of diamonds, were discovered in a later raid, but Dean Wren had more success with the registers.

By the time Christopher left school in 1646, his mother had died and Dean Wren had gone to live with his daughter, Susan, whose husband was Vicar of Bletchingdon, near Oxford. During the 3 years that elapsed between Christopher leaving school and his going up to Oxford at the age of 17, he divided his time between Bletchingdon and London. From the time when Susan Wren was married, in 1643, her husband had been instructing Christopher in mathematics, and this he applied to the design of sun-dials, a matter of some geometrical complication. Particularly complicated was the reflecting dial which Christopher made to show the time on the ceiling of a room. Amongst his other youthful inventions was a machine for use in drawing out the lines of a sun-dial on which he wrote a short treatise, and shortly after leaving school he translated from English into Latin the treatise on "geometrical dialling" which formed part of *Clavis mathematica*, by the then famous mathematician William Oughtred. This Oughtred published in 1652 with other of his works.

Another piece of Wren's work that was published at this time was a short treatise on the calculation of dates in the Julian calendar. It appeared in the 5th edition of Helvicus's *Theatrum historicum et chronologicum* (Oxford, 1651).

But Wren only spent part of his time at his sister's; for much of the period after he had left school he stayed with Sir Charles Scarburgh in London. Scarburgh was a physician and disciple of the great Harvey, but his interests were wide, and it was he who suggested the translation of Oughtred's book, and with him Wren invented the "weather clock" which was to figure prominently in his later meteorological work. Wren had first got to know Scarburgh as a patient, but was taken on by him as a pupil and later became his assistant.

Scarburgh gave lectures on human anatomy at the Surgeon's Hall in London, and on these occasions Wren was called upon to be his demonstrator. He made anatomical models in pasteboard to show the action of muscles. This was a topic which continued to interest him, and some 20 years later he discussed the action of muscles before the Royal Society. He suggested that the swelling of a muscle as it acts might be like the blowing up of a bladder, filled by air from a fermenting liquor.

Sir Charles Scarburgh was a member of the group which met to discuss natural philosophy, and which was later the nucleus from which the Royal Society grew. Other members of the group were John Wallis and Dr. Wilkins, and it may be that Sir Charles took young Wren along to some of these meetings. Certainly it was he who introduced Wren to Dr. Wilkins, a man of Puritan leanings who was shortly afterwards to become Warden of Wadham College, Oxford.

Wilkins's arrival at Oxford as a nominee of Parliament might have been the occasion of considerable internal friction, but instead, no doubt because of Wilkins's tolerant outlook, the College soon became the most notable in the University through the members it attracted, who included the sons of several Royalists. Amongst the latter was Christopher Wren, who entered the College as a gentleman commoner in 1649 at the age of 17. At about this time, other members of the London group of natural philosophers came to Oxford, again as nominees of Parliament. Their enthusiasm for natural philosophy was such that they soon began to hold meetings in Oxford like the London ones, first at William Petty's lodgings and then at Wadham College.

TRANSFUSION EXPERIMENTS

From an early date in his career at Oxford, Christopher Wren attended the meetings at Wadham as did his cousin, Matthew Wren, son of the imprisoned Bishop of Ely. Another member of the group was Thomas Willis, an eminent physician with whom Wren continued his medical studies. At various times during the next few years he did various dissections which included studies of the adder and the otter as well as the structure of fish and the anatomy of nerves. Another experiment he performed was to remove the spleen from a living dog, which survived the operation, and "in less than a Fortnight, grew not only well, but as sportive and as wanton as before".

Wren also studied the eye, and exhibited to the meetings at Wadham a model of the eye "with the Humours truly and dioptically made". Then in 1656 he proposed "the noble anatomical experiment of injecting liquors into the veins of animals". Assisting him at the first of these experiments was Robert Boyle, who later described it. Using a syringe, Wren injected a tincture of opium through a quill into the vein in a dog's hind leg, and found that the dog went into a stupor. Other drugs were tried, including wine, and Wren improved his technique until, in 1659, he tried the transfusion of blood directly from one animal to another.

Robert Hooke came up to Oxford as an undergraduate a short time after Wren, and they soon became good friends. Hooke worked for a while as Boyle's assistant, and Wren was present at some of their experiments with the air-pump. Wren himself did some experiments with barometers and was able to make useful suggestions to Boyle about the observed variations of atmospheric pressure. A less scientific activity of Wren's was his work on a new method of etching. Prints that he produced using the new process provide evidence of his ability as a draughtsman.

In 1651 Wren graduated with the degree of B.A., proceeding to the M.A. 2 years later, when he was given a fellowship at All Souls. The following year he met John Evelyn who had been invited to Wadham by his "deare and excellent friend Dr. Wilkins".

Evelyn was shown some transparent apiaries :

"built like castles and palaces, and so order'd . . . one upon another as to take the honey without destroying the bees. These were adorned with a variety of [sun] dials, little statues, vases, etc., and he was so abundantly civil on finding me pleased with them as to present me with one of the hives. . . . He had above in his lodgings and gallery variety of shadows of all perspectives, and many other artificial, mathematical and magical curiosities, a way-wiser, a thermometer, a monstrous magnet ... most of them of his own and that prodigious young scholar, Mr. Chrs. Wren, who presented me with a piece of white marble, which he had stain'd with a lively red very deepe, as beautiful as if it had been natural".

The transparent hives were probably constructed by Wren and amongst other things which Evelyn saw there may have been some of the items which Wren had demonstrated to the meetings at Wadham. These were listed in the book *Parentalia* which Wren's son compiled, and included "Divers Improvements in the Art of Husbandry" and "New Designs tending to Strength ... in Building". These would have particularly interested Evelyn, who later wrote books on horticulture and architecture. The way-wiser he saw in Dr. Wilkins's rooms was probably also of Wren's construction, for the list in *Parentalia* includes a device for measuring distances travelled by a coach along "the winding Way", and the thermometer may have been one which Wren used for his meteorological observations.

GRESHAM PROFESSOR OF ASTRONOMY

In 1657, when he was only 24, Wren was offered the professorship of astronomy at Gresham College, London. He at first declined this offer on account of his youth, but his friends were able to persuade him to change his mind.

Wren's inaugural lecture at Gresham College was a flowery piece of work, and was read in Latin. But it is a significant document in that it indicates Wren's belief in the practical value of astronomy; his admiration of Harvey and Gilbert; and his distrust of Descartes's system of natural philosophy. Descartes, he said, was "but a Builder" upon Gilbert's experiments. Even more important were his comments on Kepler, who had worked out three laws to describe the motion of the planets, using a large number of observations made by Tycho Brahe. Kepler's laws were entirely empirical, lacking any physical explanation, and Wren saw that the great problem in theoretical astronomy in his day was to find such an explanation. He went so far as to say that the solution of this problem would be worked out by someone then living. He may have hoped to do it himself, for he was working on Kepler's laws about this time.

In the next year, the French mathematician and religious thinker, Blaise Pascal, issued a challenge to other mathematicians in the form of a problem. This concerned the cycloid, which is the curve traced out by a point on the circumference of a wheel rolling along a flat surface. The mathematicians were asked to work out certain properties of the solid generated by the rotation of a cycloid before 1 October 1658, and a prize was offered for the best answer. Wren was quick to send in a solution, and accompanied it by another problem which he challenged Pascal to solve. This was a problem which had originally been raised by Kepler and concerned the properties of the ellipse. Kepler's second law, which discussed the motion of a planet in an elliptical orbit, stated that a line joining the planet to the sun would sweep out equal areas in equal times. This is because the planet moves faster when its orbit brings it nearer to the sun. Kepler's problem, now posed by Wren, was to find the speed of the planet at any point using the equal area law. Kepler had been unable to solve this problem, with the result that his law could not easily be used by astronomers, and so was ignored, even by those who used the first and third laws. Wren suggested a solution using the cycloid, and although it was not entirely satisfactory, it was a considerable advance in a field where the best mathematicians of the day had been foiled.

Another competitor for Pascal's prize was John Wallis, Wren's former teacher. His work did not please Pascal as highly as Wren's, which was judged the best although the prize was withheld. But Wallis thought his work worthy of publication, and brought it out in 1659 in a book called *Tractatus duo*. This book incorporated four theorems by Wren, including his solution of Kepler's problem, and was important because it was the first book published in England to state and discuss Kepler's second law.

At the time that this work was proceeding, Wren was trying to elucidate the problem of the planet Saturn, whose rings were first observed with a telescope a generation earlier. The problem was partly that the rings could not usually be distinguished as such; sometimes they looked like a pair of satellites orbitting the planet, and sometimes like a pair of arms. Wren described his work on the problem in a letter to Sir Paul Neile.

His views on Saturn, he says, were first

"... hatched by your Influence at White Waltham upon the observation of December, 1657 when wee first had an apprehension that the Armes of Saturn kept their length which produc'd this hypothesis [model], made first in two paste-boards, not to say anything of our attempts in Wax. . . . The hypothesis made more durable in metal was posed on the Top of that Obelliske which was erected at Gresham College in 1658. . . . To raise the 25 foot telescope of your Donation; at the same time I was put upon writing on this Subject for which I supposed I had tolerable observations and materials at Hand; . . . but when in a short while after, the hypothesis of Huygens was read over in writing, I confesse I was so fond of the neatnesse of it, and the naturall simplicity of the contrivance agreeing soe well with the physicall cause of the heavenly bodies that I loved the Invention beyond my owne and though this be soe much an equivollent with that of Huygens, that I suppose future observations will never be able to determine which is the trewist, yet I would not proceed with my designe."

WREN AND THE ROYAL SOCIETY

Wren's tenure of the astronomy professorship was interrupted by the disturbances which followed Cromwell's death in 1658. Gresham College was taken over by the army and Wren went back to Oxford. Later, after the monarchy had been restored Gresham College returned to its proper function, and Wren resumed his lecturing there.

It had been the custom for meetings of those interested in experimental philosophy to be held in Gresham College, like the ones Wren had attended at Oxford. In the autumn of 1660 these meetings were resumed, being held on Wednesdays after Wren's astronomy lecture. It was at a meeting of this kind on 28 November 1660, that it was suggested "that some course might be thought of to improve this meeting to a more regular way of debating things". Rules were drawn up, the king's patronage sought, and the Royal Society formally came into being a short time later.

Soon after this, Wren wrote a letter to the President of the Society, Lord Brouncker, reviewing what had been and what should be done. He gave a prominent place to the utilitarian value of the Society, which should aim to advance knowledge, increase profit, and improve the health and convenience of human life. He thought the Society's plans for studying industrial processes by compiling histories of trades "cannot but produce something very extraordinary".

Human health was already in the hands of an able profession, but he thought that medicine should make more use of chemistry. He wanted

"to have a fire going in the elaboratory for choicer experiments in chemistry, especially since many parts of philosophy are not to be pierced far into, without this help; and *little is to be done in the business of trades without it.** Mechanical philosophy only teaches us what probably may be done in nature by the motion and figure of the little particles of things, but chemistry helps us to determine what is actually done by the motions of those invisible parts of Liquors, Spirits and Fumes. . . . Thus in the Body of a Man, if we consider it only mechanically, we may indeed learn the Fabric and Action of the organical parts, but without chemistry, we shall be at a loss to know what Blood, Spirits and Humours are, from the due temper of which the motions of all the parts depend."

Wren here seems to be suggesting that without a greater stress on chemistry, the new "mechanical philosophy" might not be so useful as had been hoped. The passage also recalls his own suggestion of a chemical mechanism by which muscles might work. In a later letter to Lord Brouncker, Wren's interest in the chemistry of respiration is expressed. He apparently had ideas for

* Our italics.

purifying air that had been breathed, so that a man could live indefinitely in a sealed enclosure without fresh air.

"A Description of a Vessel for cooling and percolating the Air at once I ... left in Mr. Boyle's Hands ... you will at least learn thus much from it that something else in Air is requisite for Life, than that it be cool only, and free from the fuliginous Vapours and Moistures it was infected with in Expiration; for all those will in Probability be deposited in its Circulation through the Instrument. If nitrous fumes be found requisite (as I suspect) Ways may possibly be found to supply that too, by placing some benign Chymical Spirits, that by fuming may infect the Air within the Vessel."

Wren's idea that air contained "nitrous fumes" which sustained animal life was quickly taken up by Hooke, who tried to use it to explain combustion. In 1665 Hooke made experiments which were intended to show that fire consisted of the dissolution of the burning material by a "nitrous substance inherent and mixt with the air".

The records of the Royal Society show that Wren was an active member, making frequent suggestions and showing experiments. Soon after its foundation he did some experiments with pendulums before the Society, and in 1662 demonstrated the oscillations of mercury in a tube bent into a U-shape, and showed that these oscillations were of the same kind as the motion of a pendulum.

Another contribution which Wren made to the Society's early meetings concerned meteorology, a subject which had interested him since his youth. He had then made a weather clock (Fig. 3.1) by means of which the weather was recorded automatically during the night. The value of records of weather was still occupying him when he wrote to Lord Brouncker about work which the Royal Society might sponsor. Continuous records of wind, rain, temperature, pressure, and humidity over a period of years should be accompanied by a record of the growth of crops, so that the conditions for good and bad harvests might be elucidated. Careful records of epidemics should be kept, and weather reports compared with the weekly bills of mortality in London. Wren's weather clock was shown to the Society, and would have played an important part in the programme of observations which he advocated.

The instrument was driven by a pendulum clock, and pencil lines drawn on a number of moving charts recorded humidity, temperature, and barometric pressure. The idea was later taken up by Hooke, who exhibited an improved weather clock to the Royal Society in 1679.

More ingenious than the weather clock itself was the thermometer Wren made for use with it. It was bent into a circular shape and rotated as the liquid in it expanded and moved round the circumference. This instrument was an especially interesting innovation, because it provided a mechanical action which could not only be used to mark the chart of the weather clock but could also work a thermostat controlling the temperature of a furnace. Wren wanted a furnace in which steady temperatures could be maintained for long periods in order to conduct experiments on the hatching of eggs and insects, on the germination of plants, and for chemical researches.

The communication of some of this work to the Royal Society was no doubt delayed by his removal to Oxford in 1661. Seth Ward vacated the Savilian chair of astronomy to become Bishop of Salisbury, and Wren was appointed in his place. Within a short time, Cambridge University honoured the new professor by making him a Doctor of Laws, and shortly after Oxford conferred a similar degree. Wren liked Oxford, and was no doubt pleased to move back to All Souls College, although it meant that he could not attend the Royal Society so frequently. However, he continued to demonstrate experiments to the Royal Society as often as possible, and meanwhile was kept in touch by correspondence.

Wren's astronomical work at this time seems to have included a study of the moon; for a letter from Sir Robert Moray contained the king's command "to perfect the Design wherein he is told you have already made some progress to make a Globe representing accurately the figure of the Moon as the best Tube [telescope] represented; and to delineate by the Help of the Microscope the Figures of all the insects and small living creatures you can light upon, as you have done those you presented to His Majesty".

The model of the moon was completed and given to the king by Wren himself in private audience. It showed all the mountains and craters on the moon's surface and when suitably illuminated demonstrated the moon's phases. But Wren begged to be excused from the task of making drawings of insects as seen through a microscope, and Hooke did the work instead, publishing it in his famous book, *Micrographia*. He wrote that he had begun the work "... with much *Reluctancy* because I was to Follow the Footsteps of so eminent a person as *Dr. Wren*, who was the First that attempted any Thing of this Nature; whose original draughts do now make one of the Ornaments of that great Collection of Rarities in the *King's Closet*".

Wren had begun his work in microscopy by making several improvements to the instrument itself. His knowledge of optics derived mainly from Kepler, whose work in this field formed the basis of a course of lectures he gave at Gresham College. He discoursed to the Royal Society about

"a Natural and easie *Theory* of *Refraction*, which exactly answer'd every *Experiment*. He fully demonstrated all *Dioptrics* in a few Propositions, shewing not only (as in *Kepler's Dioptrics*) the common properties of *Glasses*, but the proportions by which the individual Raies cut the *Axis* . . . upon which the proportion of Eye-glasses and *Apertures* are demonstrably discover'd".

He concerned himself with the problem of grinding lenses of parabolic or hyperbolic section, possibly in the hope that such lenses would produce smaller aberrations than the conventional lenses of spherical section. In the course of this work Wren became involved in a study of the geometrical properties of the hyperboloid and discovered the family of straight lines that are contained in the surface of this solid.

FIRST WORK IN ARCHITECTURE

Wren reached the age of 30 without designing a single building, but over a short period in 1662–3 his advice was sought concerning two existing structures and he was commissioned to design two new buildings. To account for this sudden emergence of Wren as an architect one must remember that his ability as a draughtsman and model-maker was already well known, and it is probable that he had discussed architecture with his friends. Moreover, very little building work had been carried out in England during the disturbed Commonwealth period, so even if Wren's abilities as an architect had been apparent earlier, there would have been no opportunity to apply them. It is probably significant that two of these early architectural commissions involved giving advice about existing structures, not designing new ones, for in both cases it was just as appropriate to appoint a scientist as an architect. The first commission was to survey the fortifications at Tangier, which the king had recently acquired as part of the dowry of his bride, Catherine of Braganza. A letter to Christopher Wren from his cousin, Matthew, now secretary to the Lord Chancellor, conveyed this offer, but Wren declined it because he thought it would endanger his health. It has been suggested that Matthew Wren had used his influence with the king on his cousin's behalf, but it seems just as likely that John Evelyn had recommended Wren to the king. Evelyn was himself interested in architecture and would have known about Wren's abilities and interests. But it should also be remembered that the king possessed drawing by which he could judge Wren's ability. The fact that Wren refused the Tangier job cost him no disadvantage. He was shortly after appointed Assistant to the Surveyor General.

The Dean and Chapter of St. Paul's probably invited Wren to survey their cathedral on the strength of this official appointment. Extensive repairs to the building had been carried out under the supervision of Inigo Jones, but were left unfinished when the Civil War began. Since then the cathedral had suffered much neglect, even being used as a stable for a while. Wren submitted a report after an interval of some 3 years, advocating extensive reconstruction.

Two other commissions came from men who had personal contact with Wren—his uncle (the Bishop of Ely) and Gilbert Sheldon (Archbishop of Canterbury and a Fellow of All Souls until 1647). The Bishop of Ely wanted to give a chapel to Pembroke College, Cambridge. It was built between 1663 and 1665 and was Wren's first completed building. The design of the main front was adapted from a drawing in a textbook on architecture by Sebastian Serlio. But before starting on the Pembroke chapel, Wren had proved himself to his uncle by designing a doorway which was made in the north wall of Ely Cathedral.

Sheldon's commission was for a large hall in which the formal business and ceremonial of Oxford University could be performed. In drawing up his plans, Wren once again referred to Serlio's book. But in this case he seems to have based the ground plan on Serlio's reconstruction of a Roman building, the Theatre of Marcellus. This had been an auditorium open to the sky, and to adapt its design to the English climate involved giving it a roof—a considerable problem because the width required (68 feet) was much greater than an ordinary roof truss could span especially if it had also to support a flat ceiling. The roof truss which Wren eventually designed attracted wide attention because of its ingenious construction. The main structural member was a heavy beam spanning the whole width of the building and made up of relatively small pieces of timber dovetailed together and strapped up with iron bands. It is interesting to find that John Wallis had produced a geometrical theory for a flat ceiling in a wide room, but the construction that Wren used is more closely related to methods described in another architecture textbook.

The technical ingenuity of Wren's design was sufficient to make a model of it a very worthy exhibit at a Royal Society meeting in April 1663. But the decorative treatment of the building was somewhat ungainly, and demonstrates Wren's immaturity as an architect at this date. The theatre was begun in 1664 but not finished until 1669.

During 1663 Wren was absent from Oxford for a long time, surveying St. Paul's and visiting Cambridge in connection with Pembroke College chapel.

His astronomy lectures were thus neglected, and Thomas Sprat wrote to him saying that the Vice-Chancellor

"did yesterday send for me, to enquire where the Astronomy Professor was, and the Reason of his Absence, so long after the Beginning of the Term. I used all the Arguments I could for your Defence. . . . I endeavoured to persuade him that the Drawing of Lines in Sir Harry Savile's School was not altogether of so great a Concernment for the Benefit of Christendom as the Rebuilding of St. Paul's, or the Fortifying of Tangier; for I understood those were the great Works in which that extraordinary Genius of yours was judged necessary to be employed."

Another point Sprat could have made in defence of Wren was that Sir Harry (or Henry) Savile, the founder of the Savilian professorship which Wren held, was himself a man who combined interests in architecture and in science. Savile's foundation of chairs in geometry and astronomy in 1619 was very significant because previously Oxford had no permanent posts in any science but medicine. Savile's architectural interests involved him in constant supervision of building work at Oxford (Merton College and the Bodleian Library) and at Eton between 1600 and 1620. But there is a significant difference between Savile's approach to building and Wren's. Savile might choose the architectural textbooks from which his masons were to copy details and he would actively supervise their work, but he would not usually draw designs for the buildings himself. The 50 years which separated Savile and Wren saw the practice of architecture change profoundly. No longer were important buildings designed by craftsmen, but by educated men who were skilled only at the drawing board, and could not handle a hammer and chisel. Architecture was no longer only a job for artisans, but had become a pursuit combining scholarship and practical knowledge which a member of the Royal Society could properly follow.

The winter of 1664–5 was marked by the appearance of a comet, which excited Wren's attention and occupied much of his time. The motion of comets was not fully understood at this period and it was especially important to measure their motions and their distance from the earth so that their orbits could be fitted into the solar system. This Christopher Wren set out to do. He devised a geometrical method whereby "one may easily find the true parallax of a Comet, from any four exact observations of it, made at differing times in the same place"—and assuming that the motion of the earth was known. This work of Wren's was described by Hooke in one of his Cutler lectures, published in 1678. As Hooke pointed out, Wren's method was based on the assumption that the comet moved in a straight line with uniform velocity. This would only be approximately true when the comet was a long way from the sun, so Wren's computation of the comet's closest approach to the earth must have been very approximate. In February 1665 Wren reported on his observation of the comet at a meeting of the Royal Society.

PLAGUE AND FIRE

During the Great Plague which swept London in the summer of 1665, all places of assembly were shut, the Royal Society did not meet, and Wren took the opportunity of escaping the infected air and at the same time seeing Paris.

It has been said that Wren went to Paris an astronomer and returned an architect; certainly he spent his time there very actively, surveying country houses in the district round Paris as well as the new buildings in that city itself. Construction was in progress at the Louvre, and when Wren met the architect, Bernini, he was allowed a quick glance at the plans. The British Ambassador was able to introduce him to many French architects. including Mansard, who had designed extensions to the Palace of Versailles, and he collected many drawings of the buildings he had seen. These were to be a fertile source of ideas in the future. In addition, remembering the concerns of the Royal Society, he made it his business to pry into the practice of various trades in France, but his promised "Observations on the present State of Architecture, Art and Manufactures in France" seems never to have progressed very far.

When Christopher Wren returned to England in February 1666 he was probably turning over in his mind the report on the state of old St. Paul's which he was to present to the king 2 months later. His imagination had been caught by the domed churches he saw in Paris, and having decided that the central tower of St. Paul's would have to be replaced, he was determined to give the church a dome instead. Many of the remaining stone walls would have to be refaced, and he suggested that this should be done "after a good Roman manner" rather than following "the Gothick rudeness of the old Design".

These suggestions seemed somewhat radical to the Commission for the repair of St. Paul's, and there was considerable opposition from some of the members. But one of the commissioners was John Evelyn, who approved Wren's idea for the dome and helped to persuade the Commission to consider more detailed plans and an estimate for this scheme.

This meeting of the Commission had been held on 27 August 1666; 5 days later the Great Fire broke out, and raging for about 4 days, destroyed the heart of the City of London. Included in the devastation was St. Paul's, its roof burnt off and many of its walls cracked and ready to fall, as well as 86 parish churches and about 13,000 houses. This was one of the occasions of human life with which some members of the Royal Society thought they should assist, and with characteristic energy, three of them busied themselves devising plans for a new London while the ruins of the old were still smoking. John Evelyn had already interested himself in the replanning of the City. In 1662 he had sat on a commission for "reforming the buildings, ways, streets, and encumbrances and regulating the hackney coaches", and his book on smoke pollution, *Fumifugium*, had attracted the king's interest. He saw the Great

Fire as an opportunity for putting his plans into practice, and after a quick survey of the ruins carried out while they were still so hot as to burn the feet of any pedestrian, he presented his plan to the king. However, he found that Wren had got to the king before him with another plan, which was in some ways similar to his own, and, at about the same time, Hooke had drawn up a plan which he submitted to the City Corporation. But not only were these individual members of the Royal Society interested in the re-building of London; the Society itself was sufficiently concerned to reprove Wren for not showing them his plan before he took it to the king. The Society had been criticized for its failure to produce many new ideas of utilitarian value, and it was anxious to gain as much credit as possible from its association with re-planning of the City. This was a practical problem which it regarded as a perfectly proper concern of its members, and Wren's interest in it was in no way exceptional.

None of the plans submitted for the rebuilding of London was accepted, but the authors of some of them were appointed to the Commission which was to draw up regulations for the rebuilding. The king appointed three members to the Commission—Christopher Wren, and two established architects, Hugh May and Roger Pratt. Of the three members appointed by the City, Robert Hooke and Peter Mills had submitted plans, and Edward Jerman was an architect and master mason. But Pratt and May were not very active and the bulk of the work fell on Wren and Hooke, who were already good friends and formed a natural partnership. Their suggestions resulted in an Act of Parliament in 1667 which provided for the standardization of houses according to three main types, all of brick with specified floor-heights and wall-thicknesses to minimize the danger of fire. It also provided for the conversion of the Fleet River as a canal and for the provision of new quays along the Thames. And, finally, it appointed a new Commission for the rebuilding, whose members would be paid by the City Corporation. They were Hooke, Mills, and John Oliver. Of the three, Hooke was by far the most active, measuring the sites of numerous small houses before rebuilding was allowed, and supervising the widening of several streets and the improvement of the Fleet River.

Wren's greatest concern was now with the rebuilding of St. Paul's. He was convinced that the ruins left by the fire would be too weak to support the new roof which the Dean and Chapter insisted on trying to erect. His advice on this point being ignored, Wren returned to Oxford and resumed his astronomy lectures. Building work there demanded his attention, for the Sheldonian Theatre was not yet complete, and some extensions and renovations at Trinity College were now in his charge. But although he spent most of 1667 and part of 1668 in Oxford, the records of his attendance at Royal Society meetings show that he visited London quite often. At these meetings he spoke about the force of gunpowder and read a paper on the cycloid. He was also asked by the Society for his opinion on the collision of solid bodies, and he submitted a paper which was printed in the *Philosophical Transactions* along with work on the same subject by Wallis and Huygens.

THE CITY CHURCHES

In April, 1668, the Dean of St. Paul's wrote to say that attempts to patch up the ruins of that church had failed; some pillars had collapsed while repairs were in progress and revealed unsuspected faults in the masonry which still stood. Would Wren come quickly to advise them, and bring the plans he had drawn up in 1666?

Having now been proved right about the instability of the old masonry, Wren did not find it difficult to convince the authorities that a completely new cathedral would have to be built. Thus in July 1668 the king issued a warrant authorizing the demolition of the ruins, the building of at least the choir of a new church, and appointing Christopher Wren as architect.

Little is known about the first design which Wren drew up for the new cathedral except that it had a dome and a rectangular choir. But drawings of his second design survive and also a model, which was completed in 1673. By English standards it was revolutionary; instead of a long nave and choir with short transepts, the design showed a large central space roofed by a dome with four short arms projecting from it—a Greek cross plan. This was too big a departure from "cathedral-fashion" for the Dean and Chapter to be able to approve it. The large central space was acceptable as an auditorium in which sermons could be preached, but it was not felt desirable that the cathedral should be only a preaching-house. It should also have a choir with high altar and bishop's throne, and a nave for processions. So Wren was sent away to produce a new design—his third.

This gave the authorities all the traditional features they wanted. The plan was similar to that of the old church except that it was shorter and the west end had a portico very like the one designed by Inigo Jones and erected shortly before the Civil War. At the crossing Wren placed a dome which was surmounted by a pagoda-like steeple. This design was a rather unhappy compromise, but it seems to have met with approval. When it was submitted to him, in 1675, the king issued a warrant authorizing the construction to begin, Parliament having passed an Act for financing the work by a special tax on the coal brought into London.

In 1670 an Act of Parliament was passed providing for the rebuilding of churches destroyed in the fire. Fifty new buildings were required, and the work was to be paid for, like St. Paul's, by a tax on coal. Christopher Wren was appointed architect, and he was given three assistants, one of whom was his friend Robert Hooke.

Immediately after the Act had been passed, fifteen churches were begun, and by 1677, the peak year, almost thirty were at some stage of construction. Wren must soon have realized that with all this architectural work, he could no longer pretend to be very active as a mathematician and astronomer; thus in 1673 he resigned his post as Professor of Astronomy at Oxford.

Wren's fifty-three London churches (of which about half survive at the present time) are very different in character to St. Paul's. The designs are of endless variety, typical of Wren's ingenuity, and are more informal. The handling of classical detail is looser and some would say cruder, but their atmosphere is more relaxed and benign.

Summerson suggests two reasons for this difference between the cathedral and Wren's smaller churches. The first is that Wren was in some cases quite deliberately experimenting with different aesthetic and spatial effects; and the second is that he neither detailed them as fully nor supervised their erection as closely as he did with the cathedral, so that his assistants, or the masons who built the churches, had more freedom to execute the details of the design to their own taste.

Some idea of the part that Hooke played in the building of the churches can be gathered from his diary, and also from his drawings, which were greatly inferior to Wren's. Almost all of the churches under construction during the 1670's are mentioned in his diary, often in a minor connection—he arranged for bricks to be carted to the site of one and staked out the foundations of another. But six churches are mentioned very frequently and it seems that with them he exercised a special oversight.

Hooke also designed several buildings in his own right, notably the College of Physicians and the Bethlehem Hospital, both in London, and the church at Willen in Buckinghamshire. Wren submitted designs for the Monument, which marks the spot where the Great Fire started, but the structure that was finally erected, in 1670, was built to Hooke's design.

We have seen how Wren based the plan of the Sheldonian Theatre on a Roman example illustrated in Serlio's book. In planning the City churches he again referred to Roman precedent, in Serlio, and in the book by the ancient Roman architect, Vitruvius. But to point to Wren's use of Roman examples is not to deny his own inventiveness, for the ancient precedent was never slavishly followed, and many of the churches follow plans which are highly original. Amongst these, the seven "domed" churches are particularly striking, and illustrate Wren's fondness for domed ceilings or vaults.

All of these seven domes are used for their interior effect; some of them, indeed, are not structural domes at all, but are concave plaster ceilings under conventional sloping roofs. In some cases, the dome is used to give shape to an awkward site. Thus St. Benet Fink was an irregular ten-sided building, whose interior was given a hexagonal shape by the six columns which supported an elliptical dome.

St. Stephen's, Wallbrook, began in 1672, was altogether more elaborate, and is often regarded as the finest of Wren's churches. Because it was patronized by the prosperous Grocers' Company, more money was available, and it was possible to adopt an unusually ambitious plan.

St. Stephen's has a remarkably light and graceful interior, with the central dome supported by slender columns (Fig. 3.2). The dome itself is lightly constructed of wood and plaster, with an outer covering of lead, and so presented none of the structural problems which Wren had at St. Paul's. But the way in which the supports for the dome fit onto the nave and aisles can be regarded as an experimental solution of the architectural problem of St. Paul's. The dome is carried on an octagon formed by eight arches; the north, south, east, and west arches lead into the transepts, chancel, and nave respectively, but the four others run in a diagonal direction, cutting across the corners of a square. The same system is followed at St. Paul's, where the dome is again carried on an octagon formed by eight arches. But because of complications due to structural problems, the arches are here unequal in span. Wren tried to disguise this irregularity, but his treatment of the four diagonal arches was clumsy in spite of its ingenuity.

During the period that Wren was so deeply concerned with St. Paul's and the smaller churches, he had begun to settle down to a family life. In 1669, at the age of 37, he had married Faith Coghill, a native of his brother-in-law's parish at Bletchingdon. Of the two children she bore, one died in infancy, but the other, Christopher, outlived his father and compiled the records of him which were later published as *Parentalia*. But Lady Wren—as she now was, her husband having lately been knighted—did not live for long after the birth of Christopher. She died in 1675, and in the following year Sir Christopher was married again to Jane Fitz-

FIG. 3.2. St. Stephen's Church, Walbrook.

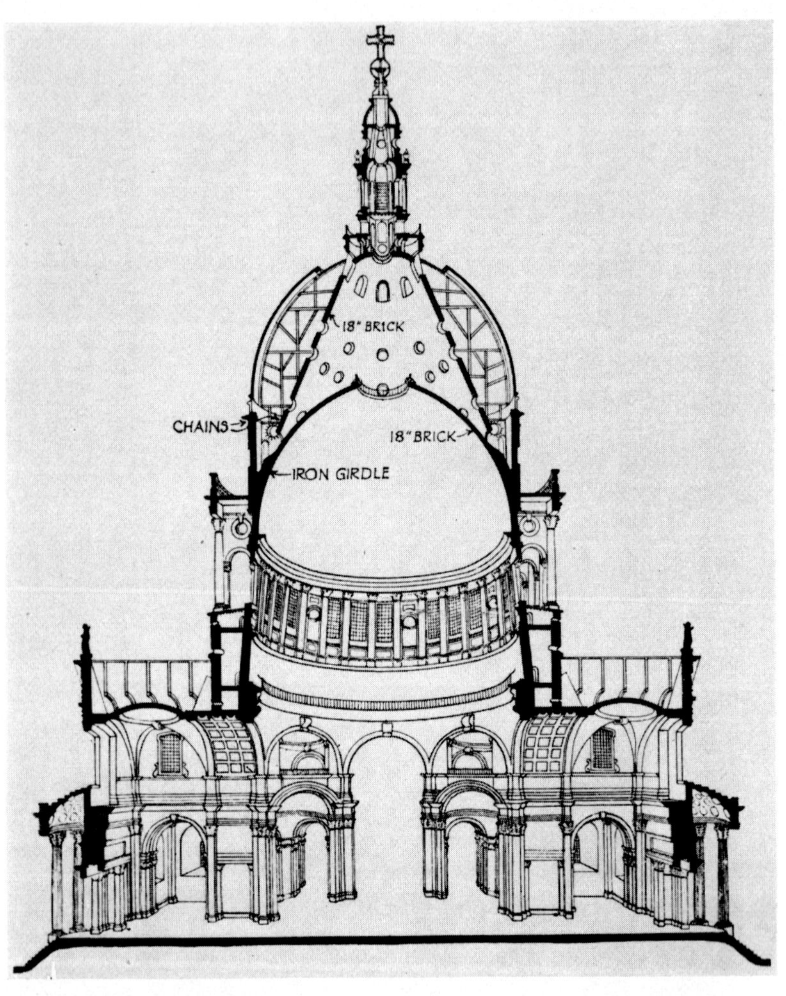

FIG. 3.3. A section of the dome and transept of St. Paul's Cathedral. (By kind permission of M. S. Briggs and T. I. Williams.)

Williams, daughter of Lord Lifford. By his second wife Sir Christopher had two more children, Jane and Stephen, but again the mother did not long survive the birth of her son. Sir Christopher did not marry again, but relied on the company of his children, his daughter keeping house for him during her short adult life.

THE LAWS OF MOTION

In 1675 the king was persuaded of the need to found an Observatory, "for the perfection of Navigation and Astronomy", and shortly after he issued a warrant instructing his workmen to build one in Greenwich Park "to such plan and design as shall be given you by Sir Christopher Wren". There is some uncertainty as to whether Sir Christopher actually drew the designs; Hooke mentions in his diary that he "described" the observatory in the company of Flamsteed and Halley, and it was he who went to Greenwich to mark out the lines of the foundations on the ground, and later to install the instruments. But this was a project in which Wren and Hooke would both have had a special interest, and probably they both contributed to its execution.

Wren's inaugural lecture at Gresham College has already been mentioned with the suggestion that he was at that time seeking for an explanation of Kepler's laws. One approach he had adopted was to study the conical pendulum, taking the circle described by its bob as an analogy of the orbit of the planet. In doing this Wren made several discoveries about the motion of pendulums, notably that the velocity of a simple pendulum bob varied as a sine function of the time. He thus discovered the periodic nature of the sine, which had hitherto been used only for simple trigonometrical applications.

Hooke was also interested in this problem and inquired into the nature of the force which caused the bob of the conical pendulum to move in a circle. He set out his views on this subject in a lecture to the Royal Society in 1666, but did not completely understand the centrifugal force until, in 1673, Huygens brought out a book on pendulum clocks. This gave an equation for centrifugal force which, with Kepler's third law, was the key to the explanation of the planetary orbits. Perhaps by this method but probably less directly, Hooke and Wren decided in 1674 that the attractive force between the sun and the planets must vary inversely as the square of the distance—the famous inverse square law of gravitation.

Newton had already discovered this law independently and so was surprised when he visited Wren in 1677 and found that Wren knew it also. Only later did Newton learn that Hooke knew the law as well, and then assumed that he had got it from Wren. But at the time, Hooke was working on the problem of planetary orbits more actively than Wren, and it seems likely that Hooke was justified in claiming that he had discovered the inverse square law and told Wren about it in 1674 or 1675.

In November 1680 the Royal Society elected Robert Boyle as its president. Boyle, however, was unwilling to serve, and Wren became president in his stead, a post which he held for 2 years. Then in 1684 the subject of planetary motion cropped up again. Edmond Halley had now worked out the inverse square law. He thought that it followed from Kepler's third law but was unable to prove it. Going up to London in January 1684 he met Wren and Hooke in an inn and discussed the matter with them, finding that Hooke claimed to be able to prove the law. But Wren was sceptical about Hooke's proof and, to encourage the inquiry, said he would give them both 2 months to bring him a convincing demonstration of the law, and offered as a prize a book worth forty shillings.

Halley apparently heard that Newton had talked to Wren about this problem; so when, 6 months later, Halley could still not find the proof he wanted, he travelled to Cambridge to see Newton. His visit resulted ultimately in the writing of Newton's great work, *Principia Mathematica* in which the laws of motion and an explanation of the orbits of the planets were fully set out for the first time. The work was completed in 1686 and printed at Halley's expense in 1687. In it Newton acknowledged that Wren and Hooke had previously known of the inverse square law, and he also referred to Wren's paper of 1668 on the collision of bodies. This had apparently contributed to Newton's ideas concerning the third law of motion, and Newton spoke of Wren, Wallis, and Huygens as "the greatest geometers of our times".

In evaluating these discussions of the laws of motion, it can be claimed that at first Wren made significant contributions which were taken up very fruitfully by Hooke, but in the 1670's and 1680's he was much more of a spectator, following with interest the work of Hooke, Halley, and Newton. Had his time not been given so fully to architecture, he might have supplied the proof of the inverse square law sought by Halley, especially as his mathematical abilities were probably superior to those of both Hooke and Halley. But it is impossible to agree with the gossip picked up by Hearne in 1726, "that Sir Isaac Newton took his famous book . . . from hints given him by the late Dr. Hook (many of whose papers cannot now be found) as well as others he received also from Sir Christopher Wren, both of whom were equally as great men as Sir Isaac". Hooke and Wren had a good physical insight into many of the problems discussed by Newton, but they lacked the vision and persistence to fuse them into a coherent system of the universe.

TOWERS AND PALACES

By 1680 the rebuilding of many of the London churches had reached the stage where construction of the towers could begin. Here there was no Roman precedent for Wren to refer to, and the variety of designs used for the towers, which form a most charming and characteristic feature of the City, must be ascribed to Wren's own inventive genius. He was, no doubt, inspired by the picturesque variety of the medieval spires which were destroyed in the fire, but most of his designs were entirely original essays in Renaissance architecture. This can be said, for example, of the spire of St. Mary-le-Bow, Wren's first grand invention of a classical steeple. But later, at St. Vedast's, the steeple built in 1697 reflects the practice of certain contemporary architects in Italy. It is an example of Wren's later period, when his use of classical forms became more baroque—that is looser and more plastic and from a certain point of view less correct—but still sober and elegant, never as sensuous in its appeal as some of the continental baroque architecture.

In 1685 Sir Christopher Wren became the Member of Parliament for Plympton St. Maurice in Devon, and later, in 1688, was elected to represent Windsor, where he built the Town Hall. He was therefore in Parliament when the question of deposing James II arose, but there is no record of which way he voted. Whatever he thought about James's replacement by William and Mary, the change was to involve him in considerable work on the royal palaces. The first job was to alter Hampton Court Palace, which was to be the principal residence of the king and queen, and work began very shortly after their accession.

Wren's first design envisaged the replacement of almost all the old palace by a grand and extensive group of buildings reminiscent of the Louvre, which he had seen in 1665. But whatever Wren envisaged, the king and queen intended something less grand, and the building eventually erected represented only an extension to the old palace, not its replacement. Perhaps, after the first design was rejected, Wren had to produce new drawings in a hurry, for the completed work is not one of his best. Its long façades are overcrowded with windows, and it seems that here Wren could produce nothing more than a stalwart, prosaic massiveness.

Work proceeded steadily at Hampton Court from 1689 until 1694, when everything was stopped as a result of the death of the queen. It was not resumed until the fire at Whitehall Palace in 1698, which destroyed the new buildings which Wren had erected there for James II. Meanwhile, Wren was working on designs for Greenwich Palace, following instructions the queen made shortly before her death that it should be turned into a Naval Hospital. As it stood when Wren began work, the palace consisted mainly of two classical buildings, by Inigo Jones and his pupil Webb, and Wren was instructed to incorporate these into his new scheme.

The site was close to the Thames, and Wren conceived a great quadrangle, open on the river side, and on the opposite side lead-

FIG. 3.4. Greenwich Hospital, now the Royal Naval College. (Crown Copyright.)

ing to a wide "avenue", flanked by colonnades. (Fig 3.4 shows one of these colonnades.) The entrance to this avenue was marked by two domes, which stood above the entrances to the hall and chapel; and for an observer in the great quadrangle, Inigo Jones's building was seen at the end of the avenue, framed by the two domes. The resulting group of buildings is one of the great masterpieces of secular architecture in England.

THE COMPLETION OF ST. PAUL'S

All this time, work was progressing steadily on the rebuilding of St. Paul's, and in 1697 the choir was ready to be used for services. Various changes had been made since the warrant design had been approved in 1675. The ground plan was not altered in any fundamental way, but the nave was shortened so that it would be the same length as the choir. This made the building more symmetrical, and was compensated for by adding a vestibule at the west end of the nave. At the same time, Wren decided to build up the aisle walls to the full height of the nave. The reason for this was partly that he wanted to disguise the nave-and-aisles plan that had been forced on him by putting the low aisles behind a high screen wall that would give the building a much more classical mass and profile. He has been much criticized for this, because the built-up aisle walls give the effect of an extra storey which is entirely false. But there were structural reasons for the walls as well as aesthetic ones.

Wren had decided that the vault of the nave needed the support of flying buttresses, and the extra height gave the walls sufficient weight to resist the buttresses' thrust. The heightened aisle walls also helped to buttress the arches which carry the dome. These changes had one disadvantage, however. The use of flying buttresses meant that the aisles had to be made slightly narrower than was previously intended; and this in turn affected the size of the arches below the dome, and resulted in the rather clumsy effect of the four diagonal arches.

These modifications must have been made quite soon after the

Warrant design had been approved, because building started in 1675 according to the revised plans. But the design for the dome and the west towers was not completely settled until very much later. The design for the west towers, completed in 1700, was derived from the same baroque sources as the tower of St. Vedast's Church, and was in some ways similar to the towers supporting the Greenwich Hospital domes, which were being designed at about the same time.

The dome at St. Paul's was started in 1697 and completed in 1709, but it presented a combination of aesthetic and technical problems which deserve detailed examination. Wren wanted the dome to stand high above his building. Old St. Paul's originally had a spire 510 feet high—the tallest building in Europe at that time—and Londoners would expect its successor to be just as much a landmark over a wide area around their city. But a very tall dome would look odd from inside the building; it would seem too much like a tall, dark, up-turned funnel. Thus from the start Wren had proposed that inside the high dome visible from the street, another dome should be built to serve as the ceiling of the church. He had seen buildings in France where this had been done successfully. The design for the high outer dome went through a number of changes and from Wren's drawings it is clear that two ideas for its shape were alternating in his mind. One idea was to have buttresses supporting the dome, like St. Peter's at Rome; the other was to have a continuous colonnade performing the same function.

The St. Peter's design eventually fell out of favour, probably for aesthetic reasons. But Wren was also aware of its technical faults. He commented that at St. Peter's ". . . the butment of the Cupola was not placed with Judgement; however, since it was hooped with iron it is safe at present". Wren had much experience of inadequately buttressed buildings like this. At Old St. Paul's he had found that the pillars were "bent outwards at least six inches from their first position", and he commonly found defects of this sort in medieval buildings because the buttresses were inadequate.

The size of buttresses needed in a given building depends on the shape of the arches and vaults which they support, so it is interesting to find that in 1671 Wren delivered a paper to the Royal Society which described the "line of an arch for supporting any weight". Hooke also had a theory as to what the ideal shape of an arch should be; this he mentioned to the Royal Society in 1671, but was very reluctant to reveal the details. When he published an announcement of his discovery in 1676 he gave an anagram, the solution of which was only found in his diary after his death. Hooke had discovered that if an arch was made to the same curve as a loop of string hanging from two points, then that arch would not require any buttresses. If the arch carried a heavy load at one point, it could be represented by hanging a weight from the corresponding point on the string. The curve of a loop of string hanging from two points at the same level is called a catenary. So when Hooke's theory came to be applied to buildings, it was said that an arch and its buttresses would be stable if it were possible to fit the appropriate catenary curve within their shape. The interesting point about all this is that it seems from Hooke's diary for June 1675 that when Wren began altering the warrant design for St. Paul's, he considered the structure of the dome in the light of Hooke's catenary curve. This was not the only time that Hooke made notes in his diary about the dome; in 1679 he went for a walk in the park with Wren, and told him of an idea for "double vaulting Paules with cramps between".

If we examine a section through the dome (Fig. 3.3) to see how it may have been affected by Wren's knowledge of the catenary, we must remember that the lantern on top of the dome was extremely heavy. Thus to represent the structure by a hanging cord, the cord must be loaded at its lowest point so that it takes up a very steep-sided catenary shape. Wren's drawings show that he had realized that the high dome must be of such a shape, although, of course, he could have arrived at this result by other means than the catenary. The problem was that a steep-sided catenary was not an acceptable shape in classical architecture. At one stage Wren thought of disguising it by means of a cladding of timber and lead that would produce the required silhouette. From that idea, he did not have to go far to reach the principle of a triple series of domes, which was his final solution—an inner masonry dome to form the ceiling of the church; an intermediate brick shell which carried the weight of the lantern; and the outer high dome, of timber and lead, which was supported by the brick shell, and shaped to give the required architectural effect. In the finished building, the brick shell, running down into the inner walls of the drum below the dome, is remarkably like an appropriately loaded catenary. If further evidence is required that it was in fact designed to a catenary shape, we may refer to a discussion between Newton and David Gregory about "Wren's problem about the loaded arch". This discussion took place in 1694 when Wren's work on the design of the dome was probably reaching its climax. Gregory mentioned "the demonstration of Newton that an upright catenary makes the strongest arch", and this demonstration probably formed the core of the paper published under Gregory's name in 1697.

At about this time Wren was asked to report on the fabric of Westminster Abbey and suggest how it might be restored. His report contained an interesting section on the buttressing of arches which could have been worked out from the catenary theory. On the other hand, a paper explaining how the size of buttresses should be calculated shows Wren adopting quite a different method. On the face of it, this method was quite unrealistic, for Wren treated the arch as if it were supported on the cantilever principle, and ignored the thrusts altogether. But from his reports on St. Paul's, Salisbury Cathedral, and Westminster Abbey we can see that he understood the thrusts of arches very well.

It seems likely that in the over-simplified theory set out in this paper Wren was trying to provide a simple rule-of-thumb method for calculating the size of buttresses needed to support an arch, and that he mistakenly omitted certain important forces.

OLD AGE

We have noted that when work on the parish churches started in 1670, Wren was given three assistants. He must always have employed a number of draughtsmen and site managers, for his work was too prolific to be sustained by one man. One of the principal of these during Wren's latter years was Nicholas Hawksmoor, who at the age of 18 had been taken on as Wren's "domestic clerk". Three years later he was giving Wren the kind of assistance that Hooke had done, and in 1691 he was put in charge of the drawing office at St. Paul's. Hawksmoor's ability was considerable, and with minor projects, like the extensions to Christ's Hospital, it was he who produced the designs, although Wren was nominally responsible. In 1699 Hawksmoor combined his work for Wren with a job as assistant to John Vanbrugh, who was then just beginning his architectural career. Between them, Hawksmoor and Vanbrugh built several large country houses, and finished the work Wren had begun at Greenwich Hospital. Acting independently, Hawksmoor designed a number of London churches as well as several college buildings at Oxford and Cambridge, and he ranks as one of Britain's greatest and most original architects after Wren.

But if in Hawksmoor Wren found a friend sympathetic to his ideas, opinion amongst those in authority was moving against him. This showed itself first in impatience about the completion of St. Paul's. In 1696 half Wren's salary as architect of the cathedral was withheld until the building was completed. This continued for the next dozen years, and, when, in 1709, the building was structurally complete, further excuses were made for delaying payment of the arrears.

Then in 1714, when George I came to the throne, Wren's difficulties increased. Although he remained nominally in office, authority was taken from him and given to the new king's nominees. These men were given charge of Wren's buildings and made alterations to his designs which were against his wishes. They disliked his architecture, preferring the new fashion of Palladianism, and they added a balustrade to the parapets of St. Paul's to make it conform to this new fashion. Then in 1718 Wren and Hawksmoor were removed entirely from their positions in the Office of Works. Hawksmoor had other building work in progress which was not affected, so his practice was not seriously interrupted. But Wren, now aged 85, retired to a house at Hampton Court Green, and returned to the interests of earlier days—he spent the last 5 years of his life working on a method for finding longitude at sea in response to the Royal Society's offer of a prize for ideas on this problem.

CONCLUSION

Christopher Wren is popularly regarded as the greatest of English architects, and certainly it would be hard to find one to surpass him, especially in church building where he reached his highest achievement. Of his relatively few domestic buildings, only Greenwich Hopital (Fig. 3.4) can compare with his finest churches. His architecture is essentially rational and is characterized by an order and clarity which is sometimes thought dry because of its lack of surprising or dramatic effects. But his interiors are tranquil and friendly, and his spires are London's most charming ornaments.

He once wrote that the beauty and strength of a building depended ". . . on geometrical Reasons of Opticks and Staticks . . . Natural [beauty] is from Geometry consisting in Uniformity (that is Equality) and Proportion . . . always the true Test of natural or geometrical Beauty". And this fairly represents his rational approach to design.

But if Wren's reputation as an architect is established, his reputation as a scientist is confused because he published so little that assessment of him is not easy. Until more research has been done, we cannot do much more than accept the valuation of his work as it was expressed by Thomas Sprat in 1667, who gave most emphasis to Wren's work on the laws of motion—pendulums, the laws of impact, and planetary motion. But Wren's interests were so diverse that he left many ideas in a tentative and inconclusive state, and it is interesting to note that some of them were taken up and explored further by Hooke. Examples are Wren's ideas on meteorology, on the composition of air, and concerning the laws of motion. With the latter, Wren can be numbered among the small group of mathematicians whose interest in Kepler provided the essential background to Newton's achievement.

As to Wren's techniques as a scientist, it is significant that wherever possible he would make a model of the object he was investigating, whether it was Saturn or the human eye. It may be that in spite of his considerable ability as a mathematician, the expression of truth in terms of figures or algebraic symbols was never entirely satisfying to him. He seems to have preferred to deal with something less abstract, that he could touch and see. This perhaps explains why he threw himself so completely into architecture—that he had a stronger feeling for the visual effects of solid three-dimensional shapes than for what could be discovered in a more abstract science. If this view is taken, the different aspects of Wren's career fall into place.

Thus in Wren's youth there was very little building going on, but the world of learning was full of excitement about the new experimental philosophy. Wren's friends and teachers were at the centre of this movement, and it was natural that he should share their enthusiasm. After the Restoration, enthusiasm for the new philosophy was probably less intense, and there was more room for other interests to assert themselves. There was also a great deal of building in progress, especially after the Great Fire, and architecture began to attract wide interest among men of learning. Thus it was natural for Wren, the draughtsman and model-maker, to move in the direction in which these practical abilities led him.

But there was no sharp break in his career; he never turned his back on science. There was, indeed, an essential unity in all the work he did, which can be summed up by saying that his interests covered all the branches of learning that were relevant to man's material environment, and that he brought the same rational but empirical approach to all the problems he dealt with, whether in astronomy, anatomy, meteorology, or building.

FURTHER READING AND REFERENCE

Birch, T., *History of the Royal Society*, 4 vols, London, 1756–7.

Bolton, A. T., *Wren Society Publications*, 20 vols, London, 1924–43.

DIRCKS, R. (Ed.), *Sir Christopher Wren*, Bicentenary Memorial Volume, London, 1923.

DOWNES, K., *Nicholas Hawksmoor*, London, 1959

'ESPINASSE, M., *Robert Hooke*, London, 1956

FURST, V., *The Architecture of Sir Christopher Wren*, London, 1956.

GUNTHER, R. T., *Early Science at Oxford*, Volumes II, III, and VIII, Oxford, 1923–1931.

HARTLEY, H. (Ed.), *The Royal Society, its Origin and Founders*, London, 1960.

JONES, H. W., 'Sir Christopher Wren and natural philosophy, with a checked list of his scientific activities', *Notes and Records of the Royal Society*, **13** (1) (1958).

MILMAN, L., *Sir Christopher Wren*, London, 1908.

ROBINSON and ADAMS (Eds.), *The Diary of Robert Hooke*, London, 1935.

SEKLER, E. F., *Wren and his Place in European Architecture*, London, 1956.

SPRAT, T., *The History of the Royal Society of London*, London, 1667.

SUMMERSON, J., *Architecture in Britain, 1530–1830*, London, 1953.

WEAVER, L., *Sir Christopher Wren, Scholar and Architect*, London, 1923.

WESTFALL, R. S. 'Hooke and the Law of Universal Gravitation', *British Journal for the History of Science*, **3** (3) (1967).

WREN, C., *Parentalia, or Memoirs of the Family of Wrens*, London, 1750.

CHAPTER 4

CHRISTIAAN HUYGENS, 1629–1695

D. E. Newbold

Christiaan Huygens was born in 1629, second son to a well-known and well-educated Dutch diplomat Constantin Huygens and his wife Susanna, daughter of a prosperous merchant. Christiaan was educated privately in both academic and aesthetic subjects, but even in his early teens showed a marked aptitude for geometry and for the construction of mechanical models. He was fortunate in his early introductions to great intellectuals of the time, notably to René Descartes who made occasional visits to Huygens's home near The Hague. Undoubtedly Descartes exerted a considerable influence on the young Huygens, not only through his *Principia* which Christiaan said "made everything in the world become clearer", but also through the emphasis which he placed on mathematics, already a subject enjoyed by the young man.

At the age of 16, Christiaan went to the University of Leyden and pursued a course of mathematics. There he was instructed by a former pupil of Descartes, and became a very keen supporter of Cartesian philosophy. But after leaving Leyden he became increasingly concerned about disagreements between Descartes's theoretical conclusions and his own scientific observations and deductions. He found these differences particularly marked in his study of the laws of collision of elastic bodies—a subject in which he became well versed and which influenced his thinking throughout his life, notably in work on the transmission of light. Huygen's concern about these disagreements is shown in correspondence between him and his former Leyden tutor and also in his later publications.

Descartes had derived his conclusions about collisions on the basic assumption that the "quantity of motion" in the universe is constant: the justification for this premiss is contained in the following extract from his *Principia philosophiae* published in 1644.

"As to the universal cause, it can be none other than God, who in the beginning created matter with its motion and rest, and who now preserves, by his simple ordinary concurrence, on the whole, the same amount of motion and rest as he originally created. For though motion is only a condition of moving matter, there yet exists in matter a definite quantity of it, which in the world at large never increases or diminishes, although in single portions it changes; namely in this way, that we must assume, in the case of the motion of a piece of matter which is moving twice as fast as another piece, but in quantity is only one half of it, that there is the same amount of motion in both, and that in the proportion as the motion of one part grows less, in the same proportion must the motion of another, equally large part grow greater. We recognize it, moreover, as a perfection of God, that He is not only in himself unchangeable, but that also his modes of operation are most rigorous and constant; so that, with the exception of the changes which indubitable experience or divine revelation offer, and which happen, as our faith or judgment show, without any change in the Creator, we are not permitted to assume any others in his works—lest inconstancy be in any way predicated of Him. Therefore, it is wholly rational to assume that God, since in the creation of matter he imparted different motions to its parts, and preserves all matter in the same way and conditions in which he created it, so he similarly preserves in it the same quantity of motion."

In this is the germ of the idea of what is now known as the "law of conservation of momentum". But there is one important difference between the idea expressed by Descartes and the modern concept of momentum: Descartes regarded "quantity of motion" as a scalar quantity, in some ways equivalent to later ideas of "energy" : from this basic principle it followed that when two bodies collided their quantity of motion should be conserved. For example, if two bodies collide in such a way that their velocities, both before and after the collision, are positive or zero with respect to a given direction, then the quantity of motion is conserved, according to this theory.

Huygens, despite strong Cartesian influences in his early life, had developed a more pragmatic approach to science than had the philosophical Descartes. He pointed out that Descartes's idea of the conservation of the quantity of motion would not be true if, for example, two equal inelastic masses collided with equal and opposite velocities and were brought to rest : their complete loss of motion is in sharp contradiction to the Cartesian rule. If, however, "quantity of motion" is considered as equivalent to "momentum", as it became known, then the law of conservation holds. Huygens's consideration of the problem of impact was not published until after his death—in a volume *De Motu corporum ex percussione* published in 1703—but a summary of his conclusions had been given in a paper to the Royal Society in 1669. His argument ran as follows.

He started by assuming that two equal elastic masses met with equal and opposite velocities v. After the impact they rebound from each other with the velocities unchanged : this follows from the symmetry of the situation and from observation that bodies can exist which are, for most practical purposes at least, perfectly elastic. Huygens now supposes that these events occur on a boat which is itself also moving with velocity v : for anyone on the boat the velocities of the masses before and after the collision are unaltered. But for anyone watching the impact from the bank, the initial velocities of the masses are $2v$ and 0, whilst the final velocities are 0 and $2v$ respectively. Had the velocity of the boat been u, then from the bank the initial velocities would have been $u+v$ and $u-v$, and the final velocities $u-v$ and $u+v$ respectively. Since u can take any value it follows that in collision equal elastic masses exchange their velocities. By the same type of argument, using the idea of a boat in motion, Huygens was able to show that the relative velocity of approach of two elastic bodies before impact is equal to the relative velocity of separation after impact: further, he extended his discussion to include collisions between bodies of different masses. He obtained what is now commonly called "the principle of conservation of linear momentum" (Fig. 4.1).

Following Galileo's work on the velocities acquired by falling bodies—the velocity gained is proportional to the square root of the vertical distance fallen through—Huygens gained the idea of kinetic energy. The word "energy" was not used at this time: the phrase *vis viva* was commonly employed by scientists of the day, especially by Leibniz who, in opposition to Descartes, regarded the true measure of the "quantity of motion" of a body as its "living force"—its *vis viva*—calculated as the product of mass and square of velocity. *Vis viva* was contrasted with *vis mortua,* the pressure exerted by a body at rest. Although the discussion about quantity of motion is best known by the Descartes–Leibniz quarrels, Huygens shows in his work that he had himself obtained a clear grasp of the essential distinction between "momentum" and "energy". Indeed, he was able to combine his ideas about *vis viva* with other work on the motion of the centre of gravity of a body to produce what is now the principle of conservation of mechanical energy.

His work on the motion of the centre of gravity of a body was also an extension of Galileo's ideas. Huygens considered the problem of the height to which a falling body could be made to rise again as a consequence only of its downward motion. He argued that if the centre of gravity of a system of particles fell and then rose to a higher level than that from which it started, then it would be possible to make weights rise, of themselves, to any desired height. Further, if the centre of gravity rose to a lesser level, then by reversing the process, once again, the weights could rise to any height. Hence it followed that the centre of gravity of the particles is able to rise only just as high as the level from which it fell.

The slowness of Huygens to publish his work provides some

CHRISTIANUS HUGENIUS

DE

MOTU CORPORUM

EX

PERCUSSIONE.

Fig. 4.1. Facsimile title page of *Motu corporum ex percussione* from *Oeuvres Complètes de Christiaan Huygens* (Société Hollandaise), Vol. XVI (Hollandsche Maatschappij der Wettenschappen).

reason for the general lack of knowledge about his substantial contribution to the development of mechanics. Further he published his arguments in geometric rather than analytic form, so that, as with Newton's *Principia*, his work presents particularly difficult problems of interpretation to later generations less well trained in this method of reasoning. His first published work *Cyclometriae* appeared when he was 22 years old. This was followed 3 years later by *De Circuli magnitudine inventa*, another mathematical work which gained Huygens an early reputation as one of the leading mathematicians of the time. But his later writings, both in mechanics and light, were not published promptly —much of his mechanics, not until after his death.

Huygens's interests were not confined to mathematics. Earlier Galileo had employed himself in a study of the planets, and had produced several detailed observations of Jupiter's moons, of the phases of Venus, and of the curious triplet which he believed to constitute the planet Saturn. Huygens considered it possible that planets other than Jupiter could possess their own satellites, but his observations of Venus and Mars yielded no such local moons. But when he studied Saturn carefully he found a star near the planet, and after repeated observation concluded that it was a satellite, known since as Titan, with a period of rotation of about 16 days.

Soon after this, when he was still only 26 years old, Huygens moved to Paris and was fortunate once again in meeting eminent men of the day, notably Herbert de Montmor, a wealthy amateur scientist who was identified with the "Montmor Academy"—an informal group of men who met regularly to discuss matters of common scientific interest, and who were the forerunners of the Académie Royale des Sciences.

Discussions at these meetings led Huygens to realize that progress in astronomy was severely limited by the design of the optical system of the telescopes of the day. Certainly the limitations were emphasized by the inability of other scientists to see Titan: Huygens attributed this to defects of the lenses and of the lens systems used in their telescopes. Further the field of view of the telescopes generally used was limited by the Galilean eye lens which was diverging (concave) : but to replace this by a converging lens, whilst enlarging the field of view, had the disadvantage of introducing very marked spherical and chromatic abberrations into the final image produced by the telescope. Huygens took advice from the best lens makers in Paris and in 1656 managed to assemble a 23-foot telescope with a magnification of about 100 times, double the power of his earlier instrument. The use of very long focus lenses, some more than 100 feet, in his telescopes enabled Huygens to reduce considerably the unwanted colour effects caused by chromatic aberration, but these telescopes, with their objective lenses on high poles, were very clumsy and awkward to use (Fig. 4.2).

For his telescopes he developed an improved design of eyepiece which consisted of two lenses. In this design the deviation of the rays of light was spread over four surfaces instead of two as in earlier instruments, with the result that spherical aberration was lessened markedly. Further, he arranged the lenses so that they were separated by the mean of their focal lengths, and thereby reduced chromatic aberration. This design of eyepiece, often still used in microscopes and known as Huygens's eyepiece, was made of two plano-convex lenses, with focal lengths in the ratio 1 :3; the stronger lens was nearer the eye of the observer and the curved sides of both faced the incident light. Sometimes this eyepiece is not considered convenient for modern use since, if crosswires are needed, they have to be fitted between the two lenses : the eyepiece requires a virtual object and cannot be used to focus on an object such as crosswires outside itself.

Huygens's discoveries about Saturn did not end with Titan. Galileo's observation that Saturn itself consisted of three bodies close together scarcely satisfied Huygens and led him to examine the planet very closely over a period of several years. By 1659 he had concluded that the correct explanation of the curious shape observed by Galileo was that Saturn was surrounded by a ring : this he described as having a solid and permanent character. This observation was a great advance on earlier work, but the division

FIG. 4.2. Telescope showing the use of very long focus lenses. Facsimile from *Oeuvres Complètes de Christiaan Huygens* (Société Hollandaise), Vol. XXI (Hollandische Maatschappij der Wetten-schappen).

of the ring into two concentric components was not seen until later in the same century. Cassini was able to make this observation when the plane of the rings was perpendicular to the line of sight of the observer; a necessary condition for the observation which was not fulfilled at the time when Huygens was carrying out most of his astronomical work. In discussing the behaviour of Saturn Huygens tried to explain *why* the ring remained in position around the planet: it is noteworthy that his explanation involved some central gravitational force, but there is no evidence that he anticipated Newton in equating this force with the radial accelerating force associated with any rotation of the ring about the planet. Huygens published his results in a small book named *Systema Saturnium*. This was received more favourably on scientific than on religious grounds, as the Church, after a period of silence, chose this moment to resume its attack, through a series of Jesuit tracts, on the Copernican system of the universe to which Huygens' work implicitly gave support.

It may have been the need for accurate timing in making astronomical observations, or perhaps the demand for reliable methods of determining longitudes at sea, which led Huygens to consider the design of pendulum clocks. In the years around 1660 there was a vigorous interest in clock design spurred on by considerable financial rewards offered in Spain and in Holland to inventors of successful methods for the finding of their longitude by ships at sea. Both Galileo and Leonardo da Vinci had earlier been involved in this quest for accurate timekeeping. Galileo's interest may have stemmed partly from his now legendary observations on the constant period of oscillation of a swinging chandelier, and also from his having used rather inconvenient constant-head water devices as clocks for his experiments with objects rolling down inclined planes. After his first design for a pendulum clock, it was suggested that Huygens had copied the work of these earlier scientists: it may be that he was indebted to others, for some features of the design since he had almost certainly seen records of their work. However there is no doubt that Huygens was the first to solve the practical problems of the design of a pendulum clock and to produce a model which could be made in quantity for general use.

Unfortunately, this clock was not satisfactory for maritime service, since the pendulum mechanism was all too easily disturbed by the motion of a ship. Huygens realized that for its successful use in a clock, the period of oscillation of the pendulum must remain constant and independent of amplitude : this condition is only fulfilled by a simple pendulum when its oscillations are of small amplitude—a requirement which it is not possible to meet on a ship at sea. His search for a method of overcoming these difficulties led him mathematically to the cycloidal pendulum whose period he deduced to be independent of the amplitude of swing (Fig. 4.3). This deduction depended considerably on much of Galileo's earlier work on acceleration, but it demanded Huygens's great mathematical ability to complete the necessary additional steps. A translation, from the *Horologium oscillatorium*, of Huygens's account of a part of his work on the cycloidal pendulum follows :

"Therefore to fix the shape of the lamina on which the solution to our problem is based, we must first determine the length of the pendulum, something which can easily be calculated from the fact that the ratio of the lengths of the pendula is equal to the ratio of the squares of the times taken for single swings. The result is that if we discover that a pendulum of length 3 feet will take 1 second exactly for one swing, one quarter of that length, or 9 inches, is necessary for one which will complete a swing in $\frac{1}{2}$ second. Likewise if we want to know the length of the pendulum of which 10,000 swings take up the space of 1 hour, this is the method of calculation : We know that a pendulum of length 3 feet goes through 3,600 individual oscillations per hour. Therefore the individual period of the oscillation of this pendulum is greater than that of the desired one by the proportion 10,000 : 3,600 or 25 : 9. Therefore as 25^2 is to 9^2—i.e. as 625 is to 81—so the length 3 feet will be to the required length, which will in fact be 4 66/100 inches.

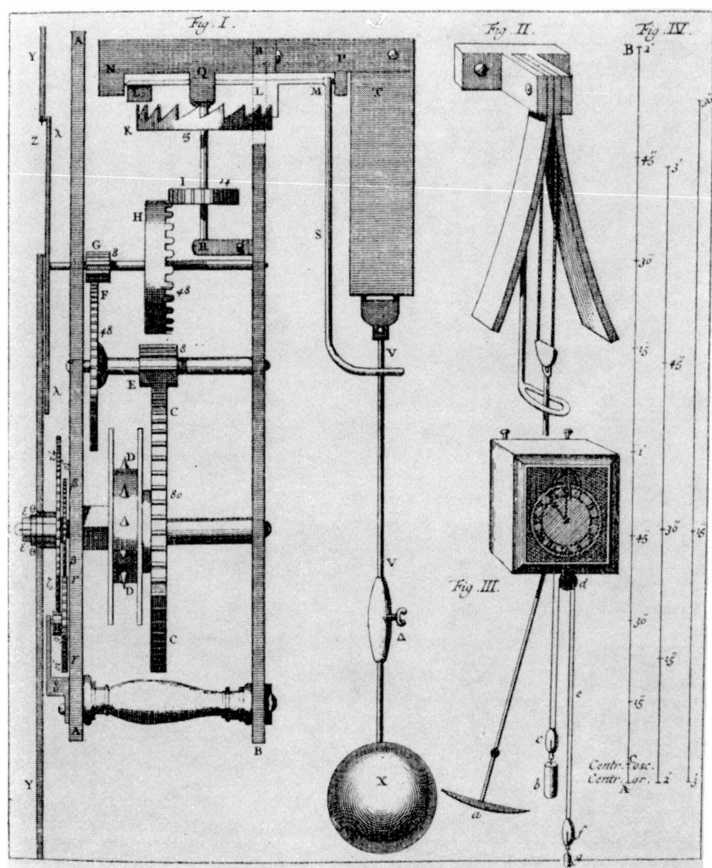

Fig. 4.3. Huygens's clock, using the cycloidal pendulum (from *Horologium Oscillatorium*). Facsimile from *Oeuvres Complètes de Christiaan Huygens* (Société Hollandaise) Vol. XVIII (Hollandsche Maatschappij der Wettenschappen).

"Once the length of your pendulum has been established, let it be 3 feet in the clock we propose—then the cycloid which is to give the curve of the guide plates can be described in this manner. On a plane surface is fixed a straight edge *AB*, about one-quarter of an inch thick. Then take a cylinder *CDE* of the same thickness, whose base has a diameter equal to half the length of the pendulum. Now take a tape *FGHE*, or better a thin metal band, and attach it to the straight edge at *F* with the other end fastened to the circumference of the cylinder at any point E, such that it is partly wrapped round the cylinder and partly stretched along the straight edge (Fig. 4.4). Then a steel stylus *DI* is attached to

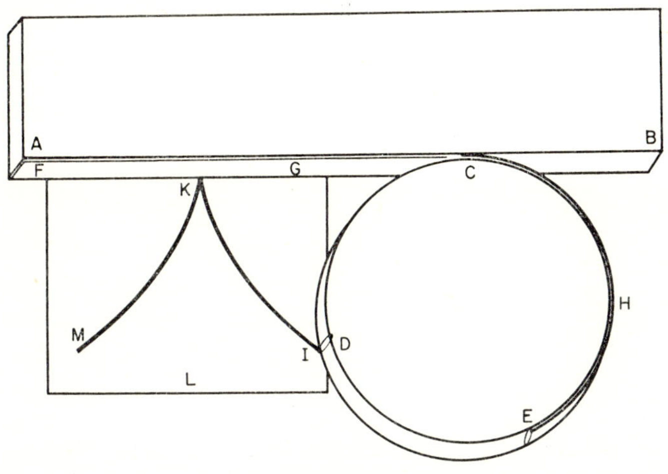

Fig. 4.4.

the cylinder extending somewhat below the base so that it reproduces exactly the position of that point on the circumference.

"Given these conditions, if the cylinder is revolved along edge *AB* with between them only the thickness of the band *FG* which is kept fully stretched, the stylus will describe a curved line *KI* which is called a cycloid, on the plane surface underneath. The circle from which it is derived is *CDE*, the base of the cylinder we are using. Now if lamina *KL* is laid alongside the edge *AB*, and if we gouge out first on it the part *KI* of the cycloid, then if we turn it over and on the opposite face gouge out a similar line *KM*, starting from the same point *K*, then we shall construct the figure *MKI* accurately on these lines, to which shape we should match the space of the guide plates between which the pendulum is suspended. Now these small parts of the curve, *KM* and *KI*, are enough for the working of the clock, since the rest of the curves which the cord of the pendulum cannot reach, would be of no use.

"In order that the remarkable properties of this line may be more fully understood, I decided to construct on another plan the complete semi-cycloids *KM*, *KI* (Fig. 4.5); these semi-cycloids

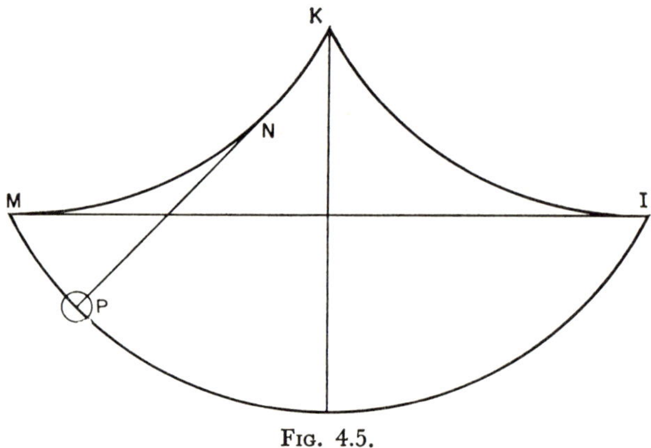

Fig. 4.5.

are such that if a pendulum *KNP* twice as long as the diameter of the circle from which they are derived is hung and set in motion between them, it will complete in a constant period of time oscillations of any amplitude up to the largest possible along the arc *MPI*. And what is more the centre *P* of the pendulum bob continually oscillates along the line *MPI* itself a perfect cycloid. I do not know whether this remarkable property, namely that it describes itself by its own evolution, has been given to any line other than this. However, what has been said will be proved in detail when I shall be discussing falling bodies and the evolution of curves.

"There is another way of constructing a cycloid by fixing points through which it passes. A circle is described with diameter *AB* equal to half the length of the pendulum. Take any number of points along its circumference at equal distances *AC, CD, DE, EF, AG, GH, HI, IK*. Join *GC, HD, IE, KF*, and the resulting lines are parallel (Fig. 4.6). Then take a straight line *LM* equal to

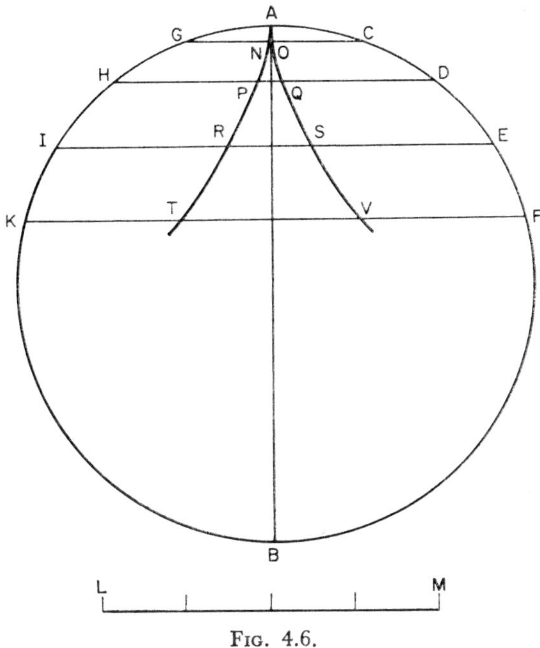

Fig. 4.6.

arc *AF* and divide it into the same number of equal parts as there are on the arc *AF*. Then construct lines equal to one of these parts, *GO, CN*, on the straight line *CG*. Let there be on the straight line *DH*, individual lines *DP* and *HQ* each equal to two portions of the straight line *LM*. Now on *EI* construct *ER, IS* each equal to three parts of *LM*, and so on if more parts have been taken. And at last when we get to the line *FK* the lengths *FT* and *KV* are both equal to the whole of *LM*. Now if curves are drawn through

AOQSV and *ANPRT*, these will be the sections of the required cycloid between which the pendulum must be hung."

In practice Huygens was able to apply the idea of the cycloidal pendulum to a clock by allowing a rigid pendulum to be attached to its pivot by a short length of ribbon which swung between curved metal plates : these plates should ideally follow the shape of the cycloidal arcs.

But despite this development and the general improvement in the standard of timekeeping which was achieved by Huygens's clocks, the basic problem of producing a clock for use at sea remained unsolved. Perhaps not surprisingly the pendulum clock, even when suspended in a ship by a ball-and-socket mechanism, still suffered greatly from the movements of the vessel and sometimes stopped completely. Although Huygens continued his researches into clock design almost up to the end of his life, he made no more real progress towards solving this problem. He heard of Hooke's attempts to produce a reliable clock based on a spring mechanism and made a brief attempt to construct a clock of this type : but he remained faithful to the idea of the pendulum clock, being certain that a spring-driven clock would be too susceptible to changes of temperature : in fact, in tests Hooke's clock proved less reliable than Huygens's earlier pendulum design.

The time and effort which Huygens devoted to work on clocks was valuable indirectly as well as directly. In the course of his researches he discovered laws governing the behaviour of the compound pendulum and the reversible pendulum, moments of inertia, and simple harmonic motion. He applied his knowledge of motion in a circle to the conical pendulum on which he based the design of another clock—but, despite some practical successes, the value of his inventions were generally far outweighed by his discoveries in mathematics.

In his work up to this time Huygens had continued to be fortunate in his acquaintance with eminent men of the day, especially through the Montmor Academy. The work of such flourishing societies aroused what may be described as amused tolerance amongst many literary figures of the day who found no interest in the "new philosophy" of science. Indeed, it may have been such attitudes which caused Montmor, in 1658, to draw up rules for the running of the Academy to try to avoid excessive time being used in purely philosophical and not very fruitful arguments—as Huygens described them. But despite difficulties, these informal societies grew, with the support in France of the chief ministers Mazarin and later Colbert, and in England of Charles II. Huygens obtained connections with the English equivalent of the Montmor Academy, the Gresham College group, and corresponded with Moray in London about various aspects of his own work. In 1661 Huygens visited London and was greatly impressed by the work of the Gresham Group : whilst there he demonstrated his own observations on Saturn and was pleased to find that his work was readily accepted. This visit to London enabled him to obtain lasting connections with the Royal Society, the direct descendant of the Gresham Group, notably through its secretary Oldenburg. So high was his regard for the English organization that some 10 years later when he was seriously ill, Huygens arranged for his scientific papers to be sent on his death to members of the Royal Society rather than to other acquaintances in Paris. Nevertheless, Huygens remained closely attached to the French society and returned to them doubtless reinvigorated by the more empirical English approach influenced as it was by Baconian philosophy.

In particular his interest in the experimental method was greatly aroused by Boyle's work, especially by the accounts given in *New Experiments Physico-Mechanicall touching the Spring of the Air* and by the *Skeptical Chymist*. Huygens reproduced many of Boyle's experiments especially those concerned with the Torricellian space above the liquid in a barometer tube. At this time the old argument dating from Aristotle about the possibility of the existence of a complete vacuum still raged. Huygens observed that the space above the liquid in a barometer tube did not always appear when the barometer was made by the method of filling and then inverting the tube into a trough of liquid. He found that if air-free water was used, the liquid could remain filling the tube even when the external pressure was reduced considerably so that the liquid level would be expected to fall : a similar result was obtained, with more difficulty, using mercury. When a small bubble of air was introduced or if the liquid contained air initially, the level of the liquid in the barometer tube fell to its expected level. This curious phenomenon, observed by Boyle also, could not be explained by the expansion of the air-bubble alone, as this was not behaving in accordance with Boyle's law. To account for his observations and to provide a medium which could explain the normal lowering of the liquid column, Huygens concluded that an ethereal substance must exist, and that this substance was capable of penetrating glass. Whilst such a conclusion has no far-reaching consequences in this part of Huygen's work, the idea of the ether is an essential element in his later work on optics where he used it in explaining refaction effects.

The advantages of Huygens's visit to London were not confined to examinations of Boyle's work or to demonstrations of his own discoveries about Saturn. His work on mechanics, especially that on momentum and energy, proved of great interest to English scientists, as they had for some years been puzzled by problems associated with the ballistic balance; his work provided the answers. It may well be that Newton derived some considerable help from this early work of Huygens : later when the *Horologium oscillatorium* was published by Huygens in 1673, it was well received in England particularly by Newton who greatly appreciated the geometrical techniques which Huygens had used. Huygens's treatment of motion in a circle and its development into the idea of "centrifugal" force was acknowledged by Newton who wrote "What Mr. Huygens has published since about centrifugal force, I suppose he had before me".

The successful launching of the Royal Society in England in 1662 with the direct support of Charles II did not provide a sufficiently powerful example for France to wish to follow immediately. There was much talk in Paris about the setting up of a similar body and of making Huygens its first director. Huygens was not in France for much of this preliminary negotiation as he had returned to Holland in 1664 and remained there for about 2 years. However, during this period his name had been kept in front of Colbert for this important appointment. Consequently when he returned to Paris in 1666 he was not only a member of the Académie Royale des Sciences when it was officially founded on 1 June of that year, but also he was designated the senior official, albeit unofficially as no formal appointment was made to such a position. Colbert gave active support to the new organization and provided accommodation for some of its early meetings, but it was not until December of that year that he persuaded the king, Louis XIV, formally to give the Académie his protection. The close relationship which existed between Huygens and leading Frenchmen, notably Colbert, proved valuable in permitting him to remain in France during much of the period of the Franco-Dutch wars, and to continue to guide the affairs of the young Académie. Huygens chose to remain outside the acute political conflicts of the times, but his continued presence in Paris, and indeed the dedication of some of his work to Louis XIV, caused a measure of consternation in his native Holland amongst fellow countrymen who failed to appreciate the very great importance which he attached to his work with the Académie Royale.

Through his residence in Paris he was able to influence greatly the development of the Académie in its early days, especially with regard to his belief in the importance of experiment and the Baconian method as exemplified by the work of the Royal Society in London. His enthusiasm for the many and varied experiments which were being performed by Hooke for the Royal Society may well have diminished further his respect for conclusions reached non-empirically and sometimes illogically by his former friend and teacher Descartes. Nevertheless, in 1669, Huygens addressed the Académie Royale and showed clearly that he still retained some elements of Cartesian influence, notably in an acceptance of the fundamental nature of the circle, and also in the use of vortex theory to account for gravitational phenomena.

Huygens was unable to accept the idea of action at a distance to explain gravitation. He pointed out that if a vessel were filled with water on top of which floated some wooden balls, then if the whole system were rotated rapidly, the balls would move towards the axis of rotation. If the wooden balls were replaced by sealing wax, which is more dense than water, then on rotation the wax would move away from the axis of rotation. But in this second case, when the rotation ceased the wax would be impelled towards the axis about which the rotation had occurred. In this idea, replacing the water by a rapidly swirling ether, Huygens saw the explanation of gravity, and found an apparently satisfactory explanation for this phenomenon. It is not easy to reconcile this part of his work with his very rigorous and logical approach to other scientific topics.

His respect for the Royal Society was equalled by their acknowledgement of his pre-eminence in several branches of science. This was shown particularly clearly in 1672 when Oldenburg sent Huygens a copy of Newton's publication on the production and nature of the constituent colours of the spectrum, and asked for an opinion of the work. Huygens appears to have misunderstood Newton's work : Newton had confined his discussion to the inherent constituent colours of white light whereas Huygens talked of the sensations produced by light in the observer. This led Huygens to argue that only yellow and blue were essential components of white light, and that all other colours were degrees of these two, since all optical sensations of colours, including white, could be obtained from them.

The ensuing correspondence with Oldenburg, and through him with Newton, must have influenced Huygens considerably, and it may not be coincidental that during the next few years Huygens undertook a profound study of optical phenomena. During these years he was intrigued also by *Experimenti crystalli islandicii disdiaclastice* written by Bartholinus about his discovery of the double refracting power of Iceland Spar. About this time also he was interested greatly by a paper on the velocity of light read to the Académie Royale by Roemer in 1676. Huygens's own work,

Traité de la lumière, was completed in 1678, but was not published until 1690.

His earlier study of the behaviour of colliding bodies led Huygens to a theory of the transmission of light : he believed that light travelled in a series of longitudinal vibrations, compressions, and rarefactions. The ether was a part of his explanations of certain phenomena associated with Torricellian space, and this now reappeared to provide a medium for the transmission of light. These ideas are shown in *Traité de la lumière*, from which these extracts are taken.

"When one takes a number of spheres of equal size, made of some very hard substance, and arranges them in a straight line, so that they touch one another, one finds, on striking with a similar sphere against the first of these spheres, that the motion passes as in an instant to the last of them, which separates itself from the row, without one's being able to perceive that the others have been stirred. And even that one which was used to strike remains motionless with them. Whence one sees that the movement passes with an extreme velocity which is greater the greater the hardness of the substance of the spheres.

". . . Now in applying this kind of movement to that which produces Light there is nothing to hinder us from estimating the particles of the ether to be of a substance as nearly approaching to perfect hardness and possessing a springiness as prompt as we choose.

". . . Also if one wishes to seek for any other way in which the movement of Light is successively communicated, one will find none which agrees better, with uniform progression, as seems to be necessary, than the property of springiness; because if this movement should grow smaller in proportion as it is shared over a greater quantity of matter, in moving away from the source of the light, it could not conserve this great velocity over great distances. But by supposing springiness in the ethereal matter, its particles will have the property of equally rapid restitution whether they are pushed strongly or feebly; and thus the propagation of Light will always go on with an equal velocity."

Huygens explains that these ether particles, although individually too small to produce observable effects, *en masse* resulted in familiar optical phenomena. This theory of lights was developed at length and has become known as Huygens's wave theory of Light : but as the above extracts show, these "waves" might be better named "pulses" as they were not originally intended to be thought of as continuous, as might be assumed from modern use of the word "wave". From these basic ideas Huygens developed the concept of secondary sources and of wave fronts to explain the transmission of light.

He started by considering the light emitted by a candle flame (Fig. 4.7). Each small region of the candle flame behaves as a

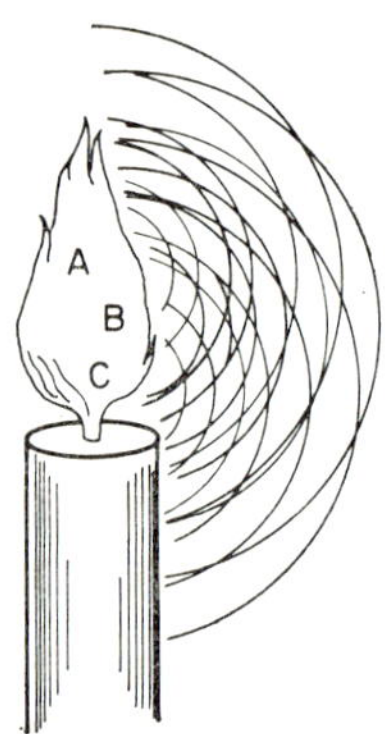

FIG. 4.7.

source of light, and sends out waves in all directions. The wave of light sent out from the whole flame is the envelope of the waves from all these component sources of light. But such a process requires the effects due to different source elements to cross each other and to do so in such a way that their separate motions are not affected by mutual interactions. It is in explaining this that Huygens returns to the concept of an ether.

"After all, this prodigious quantity of waves which traverse one another without confusion and without affecting one another must not be deemed inconceivable; it being certain that one and the same particle of matter can serve for many waves coming from different sides or even from contrary directions, not only if it is struck by blows which follow one another closely but even for those which act on it at the same instant. It can do so because the spreading of the movement is successive. This may be proved by the row of equal spheres of hard matter, spoken of before (Fig. 4.8).

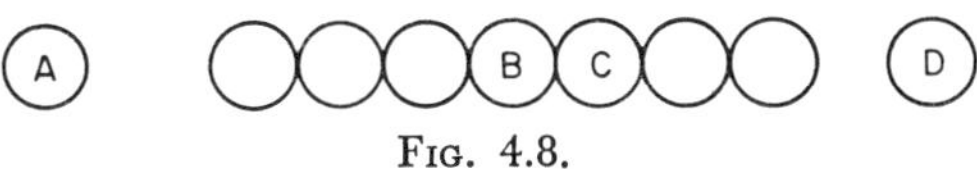

FIG. 4.8.

If against this row are pushed from two opposite sides at the same time two similar spheres *A* and *D*, one will see each of them rebound with the same velocity which it had in striking, yet the whole row will remain in its place, although the movement has passed along its whole length twice over. And if these contrary movements happen to meet one another at the middle sphere *B*, or at some other such as *C*, that sphere will yield and act as a spring at both sides, and so will serve at the same instant to transmit these two movements."

These hard particles are provided by the postulated ether; their role in transmitting vibrations is quite different from the corpuscles in Newton's theory of light where the light particles themselves move. Huygens developed the idea of the ether particle further, once again by comparison with the observed behaviour of colliding bodies. He pointed out that theoretically there was the possibility of a reflected movement as well as a transmitted movement in the transmission of light by ether particles : this possibility would be eliminated in accordance with the laws of impact if all the ether particles were assumed to be the same size.

The application of Huygens's principle to the reflection of light is well known, and is illustrated in Fig. 4.9 where a plane wave *AC* is reflected at the surface *AB* to form the plane wave *BN*. The diagrams which follow are from *Traité de la Lumière*.

Huygens remarks that his explanation of the well-known laws of reflection of light require the assumption that the velocity of the light remains constant during the reflection. He suggests that this results "from the property of bodies which act as springs; namely

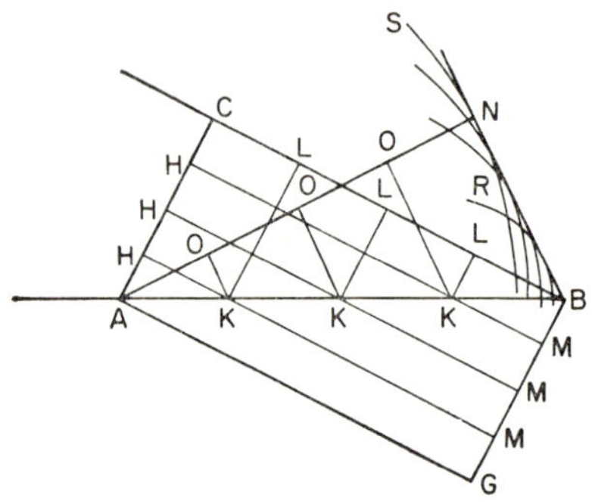

FIG. 4.9.

that whether compressed little or much they recoil in equal times". This is one more property of light explained in terms of observable mechanical phenomena, using the medium of the ether postulate.

The phenomena of optical refraction required rather more explanation than reflection. This was because Huygens had first to satisfy himself that his theory was adequate to account for the transmission of light through a solid body. He developed three possible solutions to this problem : the first depended on the light vibrations being passed on by the particles of the solid very much as they were passed on by ether particles during transmission through the air. But following his work on Torricellian space he considered it likely that the ether particles penetrated the interstices of the solid, for as he said "one may even show that these interstices occupy much more space than the coherent particles which constitute the bodies" : here the transmission of light depended on the ether particles only. In his third explanation he suggested that the transmission depended on the vibrations being passed indifferently by both the particles of the solid and by the interstitial ether particles.

The use of the particles constituting the solid, as in his explanation of light transmission through such a substance as glass, enabled Huygens to suggest a variation in the velocity of light in different media. "If the particles of transparent bodies have a recoil a little less prompt than that of the ethereal particles, which nothing hinders us from supposing, it will again be slower in the interior of such bodies than it is in the outside ethereal matter". This assumption that light travelled more slowly in glass than in air provides a means for explaining the deviation of a light wave front at a boundary between two media such as air and glass. In Fig. 4.10 a plane wave *AC* is refracted at the boundary *AB* to form

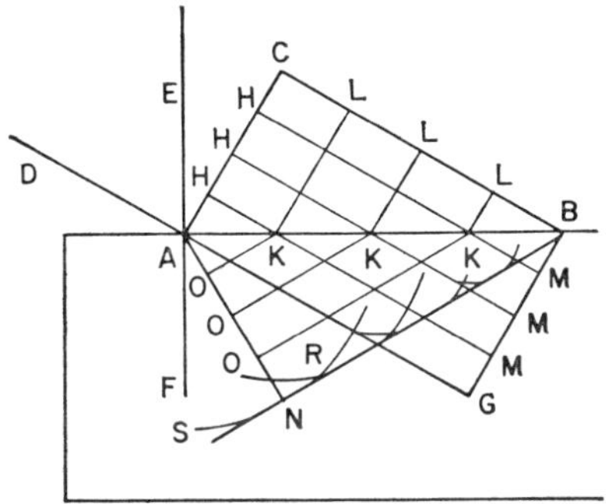

Fig. 4.10.

the plane wave *BN* : whilst a part of the wave travels from *C* to *B*, in the same time, a part travels from *A* to *N*; hence in the construction the position of the wave front after refraction is found by drawing arcs of circles in the lower medium with radii reduced from the corresponding arcs in the first medium by the ratio of *AN* to *CB* : this is the ratio of the velocities of light in the two media (and the inverse ratio of their refractive indices).

Whilst the results of this theory agreed with the observed effects of refraction, this was insufficient to validate the assumptions made in the theory, notably concerning the existence of an ether. Newton also produced an explanation of the transmission of light based on moving particles, which required the velocity of light to be increased in travelling from, say, air to glass. As at the end of the seventeenth century, there was no way known for comparing accurately the velocity of light in various media, neither theory could be discounted on the grounds of disagreement between the assumed and the observed velocities of light in various media. It was not until about a century later that Huygens's theory was shown to be the one which agreed with the observed behaviour of light in different media.

In earlier years Huygens had been somewhat scornful of Fermat's principle of least time—that the path which a ray of light takes in passing from one point to another is the path of least time. However, he discovered that this principle yielded results which agreed with his own theory of refraction, and even found it useful in explaining atmospheric refraction. He had realized that whilst in a homogeneous medium, waves from a point source would spread out spherically, in an anisotropic medium, this would not happen : the waves might spread out elliptically or in some other configuration. However, in the case of atmospheric refraction where the refractive index changes gradually, he used the least time principle to explain why "we see objects often which the rotundity of the Earth ought otherwise to hide". In his work on atmospheric refraction he made the assumption that not only the ether but also the particles of the atmosphere had some effect in determining the velocity of light : this was necessary in order to reconcile his conclusions with the observed variations in the density of the atmosphere.

About one-half of *Traité de la lumière* was concerned with explanations of the double refraction of Iceland Spar : this followed the earlier descriptive work on this topic by Bartholinus. Huygens remarks :

"In all other transparent bodies that we know there is but one sole and simple refraction; but in this substance there are two different ones. The effect is that objects seen through it, especially such as are placed right against it, appear double; and that a ray of sunlight, falling on one of its surfaces, parts itself into two rays and traverses the crystal thus."

These two rays he described as ordinary and extraordinary, the former obeying the normal laws of refraction. To account

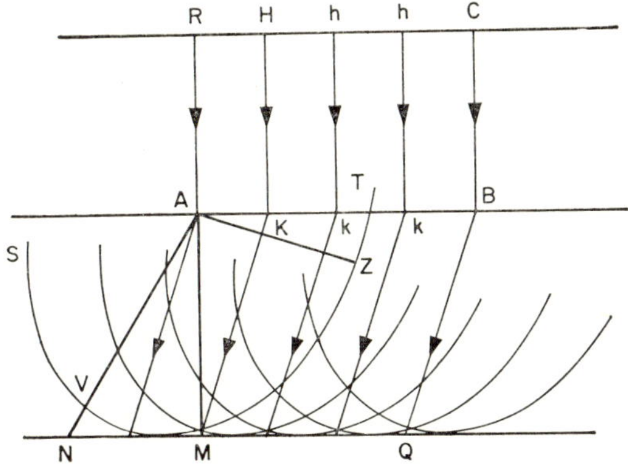

FIG. 4.11.

for the behaviour of the extraordinary ray, he assumed that this ray was propagated by a spheroidal wave motion from the point sources within the crystal: the wave front *AB* becomes *NMQ*, the common tangent of the spheroids (Fig. 4.1). This theory, whilst accounting for phenomena observed with a single crystal of Iceland Spar, did not suggest that the light itself had been altered in the two refractions. Therefore Huygens supposed that the two rays emerging from the first crystal would be split once again by a second crystal: he was surprised to find that this did not happen, but was not able to advance any satisfactory explanation. His lack of understanding of the transverse nature of the vibrations of light prevented him from developing any theory of plane polarization in accounting for double refraction, but there is no doubt that his very thorough empirical study of this phenomenon paved the way for later successful theories.

Despite his very profound study of the propagation of light, Huygens did not take up again the problem of the nature of colour after his earlier disagreement with Newton. It may be that he was not fond of such essentially qualitative work, and preferred the topics in which he could use his great mathematical ability: his training in geometry contributed very greatly to the success of his work in optics.

Over the years his health had given Huygens cause for real concern, and on several occasions he had been compelled to withdraw to his native Holland in order to recuperate. This happened again in 1681, but on this occasion he remained in Holland permanently although about 1685 he had thoughts of returning to Paris. But the political situation in France had made his position increasingly difficult, especially as the death of Colbert had deprived him of much of his former patronage. However, he did travel from Holland occasionally, and in 1689 visited England again: on this trip he attended a meeting of the Royal Society and also met Newton for the first time. Despite the divergence of their views, not only in optics but also in explanations of gravity, their personal relationship appeared amicable: each appreciated the strengths as well as the weaknesses of the work of the other.

In later years, up to his death in 1695, Huygens's work continued almost unabated. He entered into correspondence with the German mathematician Leibniz, who had developed the differential calculus. This new technique interested Huygens greatly, but although he made some use of it, he never mastered it completely. He knew about Newton's development of the method of fluxions, and played a small part as something of an intermediary in the later quarrel between these two discoverers of calculus techniques. But generally in his later years he was less concerned with highly mathematical work than in earlier years, and he turned his efforts once again to the problems of the maritime clock and of the telescope. He now reconsidered the use of reflecting telescopes on which Newton had done a considerable amount of work, but he decided that he still preferred the purely refracting type which he found to be capable of more accurate manufacture. Huygens retained the extensive literary and cultural interests which he had gained from his early education, and throughout his life showed a strong interest in the arts. He took a sceptical view of the

Church, and retained this attitude even in the last years up to his death.

With the death of Christiaan Huygens, the world lost one of the great scientists of the seventeenth century. A man of tremendous ability and intelligence, he solved many of the outstanding scientific problems of his time, sometimes losing credit for his achievements because of his slowness to publish. His range of interests is comparable with that of Hooke, but Huygens's enthusiasms were marked by depth as well as breadth. His work in mechanics, in horology, and in astronomy must not be forgotten, but it may be only just that he should be best remembered today for his theory of light of which parts are still taught in their original form some 270 years after his death.

FURTHER READING AND REFERENCE

BELL, A. E., *Christiaan Huygens.*

HUYGENS, C., *Treatise on Light* (trans. S. P. Thompson); *Complete Works of Huygens,* 22 vols. (Société Hollandaise).

MACH, E., *Science of Mechanics* (trans. T. J. McCormack).

CHAPTER 5

ROBERT HOOKE, 1635–1703

D. C. GOODMAN

IN THE middle of the seventeenth century, scientific thought was under the powerful influences of Rene Descartes, whose systematic philosophy portrayed Nature as a vast machine, and Francis Bacon, whose empiricism and utilitarianism gradually led to the foundation of the Royal Society. Of this environment, Robert Hooke was a typical product.

Born in 1635 in the Isle of Wight, Hooke was an unhealthy child who stayed at home, playing with the mechanical toys which he had himself constructed. He did not go to school until the age of 13, when he entered Westminster School. Here he rapidly mastered Euclid's *Elements* and is said to have invented 30 different ways of flying.

In 1653 Hooke went to Oxford, at that time the home of a small group of scientists who met weekly, exhibiting their experiments and calling themselves the Philosophical Society of Oxford. A prominent member of this gathering was John Wilkins, mathematician and astronomer, and a relation by marriage to Cromwell. With Wilkins, Hooke discussed his plans for flying :

"I contriv'd and made many trials about the Art of Flying in the Air, and moving very swift on the Land and Water, of which I shew'd several designs to Dr. Wilkins, then Warden of Wadham College, and at the same time made a Module, which by the help of Springs and Wings, rais'd and sustain'd itself in the Air; but finding by my own trials, and afterwards by Calculation, that the Muscles of a Man's Body were not sufficient to do anything considerable of that kind, I apply'd my divers designs whereof I shew'd also at the same time to Dr. Wilkins, but was in many of my Trials frustrated of my expectations."

A brilliant physiologist of the Oxford group, Thomas Willis, engaged Hooke as a chemical assistant and then recommended him to Robert Boyle, who had a laboratory in the High Street. Here Hooke made improvements in the air-pump, which Boyle needed for his well-known experiments on air.

Hooke's mechanical talents had so impressed the Oxford scientists that Boyle was prepared to put his name forward for a post which would bring him to the centre of English science. Since 1658 many of the Oxford group had left for London, where they joined up with a number of prominent scientists, who had for some time been meeting weekly, sometimes at Gresham College. This college had been founded in the sixteenth century by Sir Thomas Gresham, a wealthy merchant who, for the benefit of the citizens of London, had given money for professorships in divinity, astronomy, music, geometry, law, physics, and rhetoric.

In 1660 a decision was made to establish a formal scientific institution, and the members of the Gresham meetings signed an agreement to promote experimental learning. The interests of the society were comprehensive and embraced trade as well as laboratory research. Baconian in outlook, together with the mental pleasures which new knowledge afforded, the society looked to science to better man's condition through the improvement of the manual arts. In 1662 Charles II gave his patronage to this spontaneous gathering which was to be called the "Royal Society of London for the Promotion of Natural Knowledge". Without a financial grant from the king, the Society looked to its rich merchants to maintain its existence.

To carry out its experimental work, the Royal Society in 1662 created the post of Curator of Experiments, and, on Boyle's advice, Hooke was appointed. The duties of the Curator were "to furnish the society every day they met with three or four considerable experiments, expecting no recompense till the society get a stock enabling them to give it".

At first temporary, Hooke's position was made permanent with a salary of £30 and a set of rooms in Gresham College. His intensive experimentation covered a wide variety of subjects. The improvement of scientific instruments was especially emphasized by the Royal Society, and Hooke improved the telescope and invented a spring for watches. The latter brought him into an unpleasant quarrel with Huygens over the priority of discovery. Particularly delightful to the Society were Hooke's observations with the microscope. He was asked to bring one microscopical observation into every meeting, and in 1665 the Royal Society published Hooke's *Micrographia*, one of the most impressive works of seventeenth-century microscopy, outstanding for the excellence of its illustrations. More than a collection of observations, the book contained the fruits of Hooke's active and sometimes crude imagination.

It is clear from his preface to *Micrographia* that Hooke, like his contemporaries, was committed to Bacon's programme for command over Nature through experimental research. It is also clear that he shared the current conception of the world, whereby all natural phenomena had to be explained in terms of mechanics. The Cartesian mechanical explanation had replaced the barren Aristotelian occult qualities. Speaking of the use of the telescope and microscope, Hooke wrote :

"It seems not improbable, but that by these helps the subtility of the composition of Bodies, the structure of their parts, the various textures of their matter, the instruments and manner of their inward motions, and all the other possible appearances of things may come to be more fully discovered; all which the antient Peripateticks were content to comprehend in two general and (unless further explain'd) useless words of Matter and Form. From whence there may arise many admirable advantages, towards the increase of the Operative, and the Mechanick Knowledge, the which this Age seems so much inclined, because we may perhaps be inabled to discern all the secret workings of Nature, almost in the same manner as we do those that are the productions of Art, and are manged by Wheels, and Engines, and Springs, that were devised by humane Wit."

The workings of Nature were more remarkable still, according to Hooke, who believed that :

"... we shall in all things find, that Nature does not onely work Mechanically, but by such excellent and most compendious, as well as stupendious contrivances, that it were impossible for all the reason in the world to find out any contrivance to do the same thing that should have more convenient properties."

In the style of a geometrical treatise, Hooke set out his fascinating series of observations, beginning with what was most simple and gradually proceeding to the more composed so, corresponding with the point and the line, the first observations describe the appearance, under a microscope of needle points, full stops, and razor edges. Hooke observed :

"... the top of a needle (though as to the sense very sharp) appears a broad, blunt, and very irregular end; not resembling a Cone, as is imagined, but onely a piece of a tapering body, with a great part of the top remov'd, or deficient. The Points of Pins are yet more blunt, and the Points of the most curious Mathematical Instruments do very seldome arrive at so great a sharpness ... the top of a small and very sharp Needle, whose point nevertheless appear'd through the Microscope above a quarter of an inch broad, not round nor flat, but irregular and uneven; so that it seemed to have been big enough to have afforded a hundred armed Mites room enough to be rang'd by each other without endangering the breaking one anothers necks by being thrust off on either side."

The observation on the texture of cork is interesting since it contained the first use of the term "cell", which, however, Hooke invented to refer to the walls of plant-cells.

In seventeenth-century embryology the possibility of spontaneous generation was a debatable question. Though Hooke observed spores to form in moulds and saw, with the microscope, the transformation of the water gnat, he could not rule out the production of living organisms in the absence of semen. There are arguments in *Micrographia* for both spontaneous generation and seminal fertilization. The issue open to speculation, Hooke searched wildly for analogies. So the mould growing on skin book covers, and mushrooms he considered to be produced by putrefaction of animals or vegetables, by the agency of heat and moisture. The production of mushrooms was likened to the precipitation of silver from solution as a silver tree. The form of the mushroom was compared to beads of wax solidifying in a burning candle. The likeness of a stalactite to an inverted mushroom induced Hooke to discuss the "Stoppings or trillings of Lapidescent waters in Vaults underground, as a possible mechanism of the germination of mushrooms".

"Moss," said Hooke, "is a Plant that the wisest of Kings thought neither unworthy his speculation, nor his Pen."

Unable to decide how moss was generated, he sought for a satisfactory explanation of its appearance through putrefaction, and for Hooke and his contemporaries, explanation was satisfactory only if it showed how the phenomena could result from mechanics. He wrote :

"Suppose a curious piece of Clock-work, that had had several motions and contrivances in it, which, when in order, would all have mov'd in their design'd methods and Periods. We will further suppose, by some means, that this Clock comes to be broken, brused, or otherwise disordered, so that several parts of it being dislocated, are impeded, and so stand still, and not only hinder its own progressive motion, and produce not the effect which they were designed for, but because the other parts also have a dependence upon them, put a stop to their motion likewise; and so the whole Instrument becomes unserviceable, and not fit for any use. This Instrument afterwards, by some shaking and tumbling, and throwing up and down, comes to have several of its parts shaken out, and several of its curious motions, and contrivances, and particles all fallen asunder; here a Pin falls out, and there a Pillar, and here a Wheel, and there a Hammer, and a Spring, and the like, and among the rest, away falls those parts also which were brused and disorder'd, and had all this while impeded the motion of all the rest; hereupon several of those other motions that yet remain, whose springs were not quite run down, being now at liberty, begin each of them to move, thus, or thus, but quite after another method than before, there being many regulating parts and the like, fallen away and lost. Upon this, the Owner, who chances to hear and observe some of these effects, being ignorant of the Watch-makers Art, wonders what is betid his Clock, and presently imagines that some Artist has been at work, and has set his Clock in order, and made a new kind of instrument of it, but upon examining circumstances, he finds there was no such matter, but that the casual slipping out of a Pin had made several parts of his Clock fall to pieces, and that thereby the obstacle that all this while hindred his Clock, together with other useful parts were fallen out, and so his Clock was set at liberty. And upon winding up those springs again when run down, he finds his Clock to go, but quite after another manner than it was wont heretofore.

"And thus may it be perhaps in the business of Moss and Mould, and Mushrooms, and several other spontaneous kinds of vegetations, which may be caused by a vegetative principle, which was a coadjutor to the life and growth of the greater Vegetable, and was by the destroying of the life of it stopt and impeded in performing its office; but afterwards, upon a further corruption of several parts had all the while impeded it, the heat of the Sun winding up, as it were, the spring, sets it again into a vegetative motion, and this being single, and not at all regulated as it was before (when a part of that greater machine the pristine vegetable) is mov'd after quite a differing manner, and produces effects very differing from those it did before."

It was just this type of preconceived explanation, characteristic of the science of Descartes, so satisfying to the seventeenth-century mentality, that Newton fought against in order to preserve the pure results of experiment. In this battle Hooke was to be one of the adversaries with whom Newton had to contend.

The microscopic study of the water gnat inspired Hooke to write the most beautiful prose to be found in *Micrographia* :

"I could perceive, through the transparent shell, while the Animal surviv'd, several motions in the head, thorax, and belly, very distinctly of differing kinds which I may, perhaps, elsewhere endeavour more accurately to examine, and to shew of how great benefit the use of a Microscope may be used for the discovery of Nature's course in the operations perform'd in Animal bodies, by which we have the opportunity of observing her through these delicate and pellucid teguments of the bodies of insects acting according to her usual course and way, undisturbed, whereas, when we endeavour to pry into her secrets by breaking open the doors upon her and dissecting and mangling creatures whilst there is life yet within them, we find her indeed at work, but put into such disorder by the violence offer'd, as it may easily be imagin'd, how differing a thing we should find, if we could, as we can with a Microscope, in these smaller creatures, quietly peep in at the windows, without frightening her out of her usual byas."

Keeping these creatures in a glass of rain water, in which they had appeared, Hooke was surprised to see them fly away after a few weeks, leaving their husks behind. These "very strange amphibious creatures" led him to conjecture against spontaneous generation :

"Whether all those things that we suppose to be bred from corruption and putrifaction, may not be rationally suppos'd to have their origination as natural as these Gnats, who, 'tis very probable, were first dropt into this Water, in the form of Eggs, Those Seeds or Eggs must certainly be very small, which so small a creature as a Gnat yields, and therefore we need not wonder that we find not the Eggs themselves. . . ."

In 1668 a Florentine physician, Francesco Redi, experimentally disproved the spontaneous generation of worms in meat. However, the possibility of such generation in infusoria was not ruled out.

Hooke continued to interest himself in experiments on air which he had first worked on as an assistant to Boyle, and carried out important investigations into respiration. In *Micrographia* there is an interesting theory of combustion based on the model of a solvent dissolving a solid. According to this, the air contained a substance (which was also in saltpetre) which behaved as a solvent for the combustible parts of a body, whenever that body was heated in it, just as many solids would only dissolve in certain liquids by heating. In the same way that heat was often evolved during the act of solution, so the type of solution known as combustion, generated a very great heat, called fire. Other parts of the solid did not dissolve in air, but united with it to form an insoluble precipitate, either carried upwards as soot or remaining behind as sluggish particles of ash. The dissolving parts of the air being few, the combustion soon ceased, like a solution which would no longer dissolve unless a fresh supply of solvent be added. Combustion was so violent an act of solution that the air was put into a pulsating state, which constituted light.

From this, it seems that Hooke supposed combustion to be accompanied by the liberation of something to the air, as well as extraction from it. John Mayow develops this theory in 1674, calling the solvent parts of the air, which Hooke recognized also in saltpetre, as "Nitro-aerial particles". However, the correct theory of combustion was not formulated until the revolution of Lavoisier in the late eighteenth century.

Intimately connected with combustion, the phenomena of respiration had been misunderstood by William Harvey, who imagined that the purpose of breathing was to cool the blood. Boyle first carried out the fundamental experiment in respiration in 1660 using the air-pump. He placed a burning candle and a mouse in the receiver, gradually pumped out the air, and found that when the flame was extinguished, the mouse died. A few years later

Hooke applied artificial respiration to dogs and his conclusions were important advances in the physiology of respiration. Displaying the thorax of a dog, he pumped air from a bellows into a pipe inserted in the wind-pipe, and so kept the animal alive. He described this experiment to Boyle, in a letter from Gresham College, dated 10 November 1664:

"My design was to make some enquiries into the nature of respiration. But though I made some considerable discovery of the necessity of fresh air, and the motion of the lungs for the continuance of animal life, yet I could not make the least discovery in this of what I longed for, which was, to see, if I could by any means discover a passage of the air out of the lungs into either the vessels or the heart. . . . I observed this, that at any time, if the bellows were suffered to rest, and that by that means the reciprocal motion of the Lungs were not continued, the animal would presently begin to die, the lungs falling flaccid, and the convulsive motions immediately seizing the heart and all the other parts of its body; but upon the renewing the reciprocal motions of the lungs, the heart would beat again as regularly as before, and the convulsive motions of the limbs would cease."

For the first time this experiment proved that the movements of the chest were not essential to respiration and that they only provided one means for collapsing and expanding the lungs. The important conclusion that even the lung motions were only incidental to respiration had not been reached by Hooke at the time of the above letter to Boyle. But in 1667 he found that an animal could be kept alive without any motion of the lungs and that the really essential requirement was a continued supply of fresh air. Hooke conceived that the purpose of the air was to remove the fumes of the blood. However, less than a year later he pointed out that dark coloured blood, exposed to the air, soon became florid, and suggested an experiment to see if blood, passing out of the lungs, had the same floridness, which would indicate a mixing of air with the blood in the lungs.

Richard Lower, a physiologist who had worked with Hooke on

the artificial respiration of dogs, pursued this hint and proved that the difference in colours of venous and arterial blood was caused by the exposure of venous blood in the lungs, to air. Maintaining an animal by artificial respiration, Lower observed the blood in the pulmonary vein to be florid; but it was dark and venous when the supply of air was stopped. Mayow advanced further when he asserted that the blood, during respiration, combined with only a part of the air. However, further advances in the physiology of respiration were not made until the chemists of the next century.

The ninth observation of *Micrographia* announced a series of optical phenomena which were to have profound effects on the seventeenth-century conception of light. For here was the first systematic study of the colours produced by thin plates, indicative of the periodic nature of light. Working with Muscovy glass (mica), a flaky, transparent material, Hooke observed :

"... this stone has a property, which in respect of the Microscope, is more notable and that is, that it exhibits several appearances of Colours, both to the naked Eye, but much more conspicuously to the Microscope, for the exhibiting of which, I took a piece of Muscovy-glass, and splitting or cleaving it into thin Plates, I found that up and down in several parts of them I could plainly perceive several white specks or flaws, and others diversely coloured with all the Colours of the Rainbow; and with the microscope I could perceive, that these Colours were ranged in rings that incompassed the white speck or flaw, and were round or irregular, according to the shape of the spot which they terminated, and the position of Colours, in respect of one another, was the very same as in the Rainbow."

The same colours were found in thin-blown glass and in bubbles of various solutions. Investigating the conditions necessary for the coloured rings to appear, Hooke pressed two small pieces of ground-glass together, first dry and then moistened with water. The intermediate air or water film was sufficient to produce the colours. The correct conclusion was drawn that the essential requirement was a thin transparent medium, placed in a reflecting medium of different refractive index, or between two different reflecting media, each of refractive index different from the enclosed film. Hooke noticed that the transparent plate could be non-uniform in thickness, but that there were certain upper and lower limits, between which this thickness had to lie for colours to occur. He made no quantitative determinations relating the thickness of the film to the colour, and this was left for Newton to accomplish, using the superior technique of a spherical surface in contact with a plane. However, Hooke did realize that the cause of coloured rings was some combination of light reflected from both parts of the bordering media. This conclusion was the result of an experiment in which an opaque mercury film, pressed between two glass plates, was devoid of colour.

The hypothesis by which Hooke sought to explain this, and indeed all other phenomena of colours, was a wave theory, so vague as to go beyond the intelligible. He regarded luminous bodies as objects having their parts in violent motion and reasoned as follows :

"That the shining of sea-water proceeds from the same cause may be argued from this, That it shines not till either it be beaten against a Rock, or be some other wayes broken, or agitated by Storms, or Oars, or other percussing bodies. And that the Animal Energyes or Spirituous agil parts are very active in Cats eyes when they shine seems evident enough, because their eyes never shine but when they look very intensly either to find their prey, or being hunted in a dark room, when they seek after their adversary, or to find a way to escape. And the like may be said of the shining Bellies of Gloworms, since 'tis evident they can at pleasure either increase or extinguish that Radiation."

But the diamond, in particular, induced Hooke to explain luminosity by a rapid motion of particles, which, moreover, was vibratory. For a diamond, rubbed in the dark, continued to shine for some while after, and he said :

"if the motion of the parts did not return the Diamond must after many rubbings decay and be wasted; but we have no reason to suspect the letter, especially if we consider the exceeding difficulty that is found in cutting or wearing away a Diamond."

These short vibrations, he supposed communicated to the eye, from the object, with "unimaginable celerity" by an intervening medium, which following Descartes, consisted of a rarefied fluid. In homogenous media, the motion that was light was propagated in straight lines, with equal velocity in all directions, like rays from the centre of a sphere, or in the way that waves on water spread into greater circles from the point where a stone had fallen in. So long as light travelled in homogenous media, the waves were propagated at right angles to their line of direction. When light passed obliquely from one medium to another, however, the former uniform motion was disturbed and, he said : "That, which was before a line, now becomes a triangular superficies, in which the pulse is not propagated at right angles with its line of direction, but ascew."

As the oblique waves entered the unmoved medium, that part of the pulsation which went ahead was slowed down by the resistance of the medium, while the following part, having its way prepared, was less impeded. Hooke concluded : "That Blue is an impression on the Retina of an oblique and confus'd pulse of Light, whose weakest part precedes, and whose strongest follows. And that Red is an impression on the Retina of an oblique and confus'd pulse of light, whose strongest part precedes, and whose weakest follows."

So, for Hooke, as well as for all scientists before Newton, light was naturally simple and white, and appeared coloured only when it was put into a disturbed condition. Moreover, Hooke asserted that there were only two primary colours—red and blue—from whose mixture all the other colours were produced.

Soon after the publication of this theory, Isaac Newton had bought himself a triangular glass prism, and in 1672 he sent a letter to the Royal Society detailing the results of his experiments. Contrary to what has been previously supposed, Newton demonstrated that white light was composed of rays of different refranibility and that there were seven primary colours originally in white light.

This paper became the source of a controversy between Newton and his contemporaries. The discussion transcended the interpretation of light and became a debate on the methodology of science. In essence, Newton's Cartesian contemporaries were searching for hypotheses which showed natural phenomena to be the results of mechanical causes. Newton tried to find the ultimate causes of phenomena, which he also believed to be mechanical, but insisted that this was a gradual research, and could only be approached through the immediate task of investigating the laws of Nature by experiment. Certainly Newton would not allow any preconceived theories to interfere with the results of such experimentation.

So it came about, that what Newton had presented as a factual account of white light, discovered by experiment, was taken to be a mechanical hypothesis, and discussed as such. A criticism in this spirit came from Hooke, who reasserted his own theory, claiming that it explained optical phenomena just as well as Newton, and with more simplicity. He argued :

"But . . . that there are an indefinate variety of primary or original colours, amongst which are yellow, green, violet, purple, orange, etc. and an indefinate number of intermediate graduations. I cannot assent thereunto, as supposing it wholly useless to multiply entities without necessity, since I have elsewhere shewn, that all the varieties of colours in the world may be made of two."

Further, the statement that white light was originally composed seemed improbable to Hooke :

"But why there is a necessity, that all those motions, or whatever else it be that makes colours, should be originally in the simple rays of light, I do not yet understand the necessity of, no more than that all those sounds must be in the air of the bellows, which are afterwards heard to iffuse from the organ-pipes; or in the string, which are afterwards, by different stoppings and strikings produced, which string (by the way) is a pretty representation of the shape of a refracted ray to the eye; and the manner of it may be somewhat imagined by the similitude thereof : for the ray is like the string, strained between the luminous object and the eye, and the stop or fingers is like the refracting surface, on the one side of which the string hath no motion, on the other a vibrating one. Now we may say indeed and imagine, that the rest or streightness of the string is caused by the cessation of motions, or coalition of all vibrations; and that all the vibrations are dormant in it : but yet it seems more natural to me to imagine it the other way."

Consequently, Hooke concluded :

"I do not therefore see any absolute necessity to believe his theory demonstrated since I can assure Mr. Newton, I cannot only solve all the phaenomena of Light and Colours by the hypothesis I have formerly printed, and now explicate them by, but by two or three other very differing from it and from this, which he hath described in his ingenious discourse."

So Newton was confronted with a challenge to demonstrate his so-called theory. The reply to Hooke was in the same vein as that to Huygens and the other Cartesians. The experimental discovery had to be taken independently of any hypothesis. Hooke's criticism had surprised Newton, who wrote : "I find the Considerer somewhat more concern'd for an Hypothesis, than I expected. . . ."

Of the claims which Hooke had made for the explanatory power of his own theory, Newton commented firmly : "But whatever be the advantages or disadvantages of the Hypothesis, I hope I may be excused from taking it up, since I do not think it needful to explicate my Doctrine by any Hypothesis at all."

Nevertheless, his opponents forced Newton to speculate on the mechanical causes of optical phenomena, a course which he was far from being reluctant to pursue, provided this left him free to investigate the laws of Nature by experiment. But the sensitive disposition of Newton could not withstand controversy and it was no coincidence that his treatise *Opticks* was not published until 1704, the year after the death of Hooke. In this work Newton made his stand clear from the outset: "My design in the Book is not to explain the Properties of Light by Hypothesis, but to propose and prove them by Reason and Experiments."

More than in any other part of his work, the talents and inadequacies of Hooke as a scientist were displayed in the treatment of the central problem of the time, that of the planetary motions, finally solved by Newton.

Hooke's early ideas on this subject show the impress of the prevailing cosmology of Descartes. According to this, the heavens were filled with finely divided, sand-like particles, which, whirling in vortices, carried the planets along their observed courses. Like the action of a vast centrifuge, the planets settled into their orbits at various distances from the central sun. In 1663 Hooke described an experiment on the density of hot and cold water, which he considered useful to the discussion of celestial physics:

"A small bolt-head of glass being, by water put into the hollow of it, so poised, that it was but a little lighter than water, was afterwards, at the small end, sealed up hermetically. Then being put into a glass of cold pump water, it remained suspended at the top of the water; and being thrust down to the bottom, it would of itself quickly ascend again to the top, and there remain. This glass of water being set by the fire, whereby the water began to be warm, the ball within the space of a minute, began to descend, and so continued, till it came to the middle of the glass; where (the glass being at that time removed from the fire, and placed upon a table in the room) it remained suspended. So that if it were by a stick depressed below, or raised above, that middle place of the glass, it would, being let alone, return to it again, till the water again growing colder, it began to reascend to the top of the water, the place from whence it at first descended."

From this, Hooke conjectured:

"For to speak hypothetically, it may be supposed, that the vast space of the vortex of the sun or the heavens, wherein the sun, earth, and planets are contained and moved, may be filled with a fluid body, whose parts are of different densities, according as they are nearer or farther from the great fire of the world, the sun, which may be placed in or near the centre of that space, according to the Copernican hypothesis. Next it may be supposed, that the several bodies of the planets and earth may be hollow like so many glass bubbles; and though they appear much more massy than the ambient ether, they may, perhaps, as to their whole bulk, be in an aequipondium to the ambient fluid. And so, according as they are more or less massy, they may take their several stations in the fluid ether."

However, this affiliation to Descartes was short-lived and Hooke soon began to look elsewhere for inspiration. The first signs of this change appeared in the final paragraph of *Micrographia*. Descartes had explained terrestrial gravitation by the surrounding vortex of matter, which rotated the earth daily, and forced down heavy bodies, which were separated from the earth's surface. Hooke, by a traditional argument, maintained that the spherical shape of the moon showed that gravitation existed there, and further objected that since the moon did not rotate about its centre, the Cartesian account of local gravitation could not be correct. He argued :

". . . it being very probable, that the Moon has a principle of gravitation it affords an excellent distinguishing Instance in the search after the cause of gravitation, or attraction, to hint, that it does not depend upon the diurnal or turbinated motion of the Earth, as some have inconsiderately supposed and affirmed it to do; for if the Moon has an attractive principle, whereby it is not only shap'd round, but does firmly contain and hold all its parts united, though many of them seem as loose as the sand on the Earth, and that the Moon is not mov'd about its center; then certainly the turbination cannot be the cause of the attraction of the Earth; and therefore some other principle must be thought of, that will agree."

It will be shown later that the cause of gravity, which Hooke finally adopted, referred to a different type of motion of the earth.

In 1666, turning aside from Cartesian considerations, Hooke asserted that a planetary orbit was the resultant of a central attraction on a Linear motion. In the same year as Hooke, Borelli also realized the need for a central force, which would prevent the planets escaping from the solar system, at a tangent. However, Borelli regarded this central force as the expression of the natural appetites of the planets from the sun, which itself was unaffected. At the same time, in Cambridge, Newton was working on the same problems with mathematical rigour; but of this the world then knew nothing.

Hooke's analysis, which was given in terms of the conical pendulum, was accompanied by experiments to demonstrate the compounds nature of elliptical motion :

"For which purpose there was a pendulum fastened to the roof of the room with a large wooden ball of lignum vitae on the end of it. And it was found that if the impetus of the endeavour by the tangent at the first setting out was stronger than the endeavour to the centre, there was then generated an elliptical motion, whose longest diameter was parallel to the direct endeavour of the body in the first point of impulse. But if that impetus was weaker than the endeavour to the centre, there was generated such an elliptical motion, whose shorter diameter was parallel to the direct endeavour of the body in the first point of impulse. And if they were both equal, there was made a perfect circular motion."

In 1674 Hooke read his first paper as Cutlerian Lecturer, a position given to him by Sir John Cutler, with a salary of £50. Hooke continued to lecture in spite of Cutler's refusal to pay the agreed amount, and it was not till 1696, 3 years after Cutler's death, that he finally received the money owing to him. The first Cutlerian Lecture, entitled "An Attempt to prove the Motion of the Earth by Observations", was remarkable for its containing the first published statement of a universal gravitation. Promising at some future date to present a new mechanical world system, Hooke stated that this would be based on three suppositions :

"First, that all Celestial Bodies whatsoever, have an attraction or gravitating power towards their own Centers, whereby they attract not only their own parts, and keep them from flying from them, as we may observe the Earth to do, but that they do also attract all the other Celestial Bodies that are within the sphere of their activity; and consequently that not only the Sun and Moon have an influence upon the body and motion of the Earth, and the Earth upon them, but that Mercury also Venus, Mars, Jupiter and Saturn by their attractive powers, have a considerable influence upon its motion as in the same manner the corresponding attractive power of the Earth hath a considerable influence upon every one of their motions also. The second supposition is this that all bodies whatsoever that are put into a direct and simple motion, will so continue to move forward in a streight line, till they are by some other effectual powers deflected and bent into a Motion, describing a Circle, Ellipsis or some other more compounded Curve Line. The third supposition is, that these attractive powers are so much the more powerful in operating, by how much the nearer the body wrought upon is to their own Centers. Now what these several degrees are I have not yet experimentally verified; but it is a notion, which if fully prosecuted as it ought to be, will mightily assist the Astronomer to reduce all the Celestial Motions to a certain rule, which I doubt will never be done without it."

In 1679, Hooke, as Secretary of the Royal Society, began a correspondence with Newton, who had withdrawn from the scientific arena to contemplate in privacy. Hooke asked Newton for the solution of a traditional problem : to trace the path of a heavy body falling towards the centre of a rotating earth. In the course of this discussion, in 1680, Hooke stated that the force of gravity was inversely proportional to the square of the distance, measured from the centre of the gravitating mass. Certain errors in Newton's replies Hooke chose to correct in public, and the correspondence became strained. However, in being forced to consider problems which he had put aside since 1666, Newton at last deduced Kepler's laws from an assumed inverse square law of distance; but once more kept his results to himself.

Hooke put forward his mature ideas on gravity in 1682 in a paper to the Royal Society, entitled "A Discourse of the Nature of Comets". In this he described experiments on the variation of gravitation with distance, restated the inverse square law and speculated curiously on the cause of gravity. Of his experiments, Hooke wrote :

"I cannot find by any certain Experiment, that grave Bodies do sensibly decrease in Gravity, tho' further removed from the Surface of the Earth; which was the Intent of an Experiment I formerly tryed at the top of the Steeple of St. Paul's and at Westminster Abby, and may now again be repeated with much more conveniency and greater advantage at the Column on Fishstreet Hill. For by counterpoising two Weights in a curious Pair of Scales, first at the top of the Steeple, and then letting down one of the Weights by a Wire of two hundred and four Foot in length, the Counterpoise remaining at the top in the Scale, the Aequipondium remained, whereas if the Gravity of the Body had increased by Approximation to the Earth, the Weight let down to the bottom must have weighed the heavier. But though the Difference were insensible in so small an height, yet I am apt to think some Difference may be discovered in greater heights, and by some more curious ways than those I then used, even in that height."

Proceeding to the discussion of the cause of gravity, Hooke saw a true weakness in the Cartesian account. The vortex which Descartes supposed to rotate the earth daily and to thrust heavy bodies to the earth's surface, would in fact bring bodies, not to the centre of the earth, but to a line of latitude. Hooke wrote : "The Cartesian doctrine, and that of Mr. Hobbs are both insufficient, because they do not give any reason why Bodies should descend towards the Center under or near the poles."

The theory preferred by Hooke was less of a complete rejection than a refinement of Descartes. Scanning natural phenomena for analogies to gravitation, Hooke evoked the traditional comparisons with magnetism and electrical attraction, including Newton's illustration, whereby small pieces of paper were drawn up to a piece of glass that had been rubbed. Since electrics were excited by a rubbing motion, Hooke attributed their activity to a motion of their internal parts. Such motion he regarded as the first stage in the process of communication of actions to a distance. Already in optics he had supposed luminous bodies to transmit the violent vibrations of their particles to the eye by means of the intervening etherial medium. The sound of a bell reached the ear by a similar process. Convinced that he had at last found the true cause of gravity, Hooke wrote :

"I conceive then, that the Gravity of the Earth may be caused by some internal Motion of the internal or central Parts of the Earth; which internal and central Motion may be caused, generated and maintained by the Motion of the external and all the intermediate Parts of its Body : so that the whole Globe of the Earth may contribute to this Motion, as it will happen to a Globe of Glass or solid Metal, to any part of which no internal Motion can be communicated, without at the same time affecting the whole with the same Motion."

To demonstrate how action at a distance might occur, Hooke had 11 years before presented an experiment which he considered useful in the search for the cause of gravity :

"One [experiment] contrived by Mr. Hooke, whereby some flour put in a wide shallow glass, with a large sloping brim and a pretty tall foot, was made to rise and run over like a fluid, by the knocking of the glass, and by the forcible moving of one's finger round about the upper edge of the same. Leaden bullets likewise being put in this glass moved in it like a fluid on its being knocked."

A similar experiment was performed at a later date :

"The experiment was very considerable, though plain, giving a further explanation of gravity, by making a large glass vibrate, with a viol bow: by which vibration, a certain undulation is plainly seen to dart out from all such places where the glass vibrates. And it was very plainly visible, that the water, and bodies in it, did move towards every such vibrating part, and from every other part that was at rest."

Exactly this process, Hooke supposed to be taking place in the ether, that medium of communication, consisting of matter so rarefied that it escaped the senses. Describing the mechanism of falling bodies, he said:

"Suppose then that there is in the Ball of the Earth such a Motion, as I, for distinction sake, will call a Globular Motion, whereby all the Parts thereof have a Vibration towards and fromwards the Center, or of Expansion and Contraction; and that this vibrative Motion is very short and very quick, as it is in all very hard and very compact Bodies: That this vibrative Motion does communicate or produce a Motion in a certain Part of the Aether, which is interspersed between these solid vibrating Parts; which communicated Motion does cause this interspersed Fluid to vibrate every way in Orbem, from and towards the Centre, in Lines radiating from the same. By which radiating Vibration of this exceeding Fluid, and yet exceeding dense Matter, not only all the Parts of the Earth are carried or forced down towards the Center; but the Motion being continued into the Aether, interspersed between the Air and other kinds of Fluids, it causeth those also to have a tendency towards the Center; and much more any sensible Body whatsoever, that is anywhere placed in the Air, or above it, though at a vast Distance; which Distance I shall afterwards determine, and shew what proportioned Power it acts upon Bodies at all Distances both without and within the Earth: For this Power propagated, as I shall then shew, does continually diminish according as the Orb of Propagation does continually increase, as we find the Propagation of the Media of Light and Sound also to do; as also the Propagation of Undulation upon the Superficies of Water. And from hence I conceive the Power thereof to be always reciprocal to the Area or Superficies of the Orb of Propagation, that is duplicate of the Distance, and will plainly follow and appear from the consideration of the Nature thereof, and will hereafter be more plainly evinced by the Effects it causes at such several Distances."

This account of Hooke has a resemblance to the Cartesian explanation; but there are clear differences. Whereas Descartes referred terrestrial descent to an ether which whirled and carried the earth along with it, Hooke gave the original motion to the particles of the earth, and it was only as a result of this motion that the ether moved and forced bodies downwards to the earth's surface. What this motion of the earth particles was Hooke did not explain; certainly he did not mean the daily rotation of the earth:

"For if the Principle or primary Cause of Gravity, which I conceive an internal Motion in the Earth, be every way uniform, and so cause an equal Attraction to the Center, then any other Cause that alters the Dispositions of Bodies to receive this Power, or that superinduces another Power that in some parts of the Earth has a greater Renitency against the Power of Gravity than it hath in other Parts, then the uniform Effect which Gravity alone would operate, will be altered by the adventitious Power. Now the Diurnal Rotation of the Earth doth superinduce such a Power...."

Differing again from Descartes, who distributed matter in the ethereal vortex in terms of its tendency to recede from the centre, Hooke gave to his ether a vibrating, breathing-like motion, which returned heavy bodies to the earth. Following Descartes, Hooke supposed the ether to consist of particles which were finely divided, but having great density. Such an ether Newton later cleared away from the heavens, leaving only subtle vapours which would not impede planetary progress. Unlike Descartes and Newton, Hooke was not certain that the ether would do equally well for the propagation of light and sound and for gravitational attraction; at least there is a blatant contradiction in Hooke's paper. For, on the one hand, he wrote:

"That this exceedingly or indefinitely fluid Medium, though it do not at all, or at most very little, hinder the Motion of Bodies through it, is yet notwithstanding the Medium by which the Communication of the harmonious Motions of the more solid Parts and Particles are communicated to others at a considerable distance; and that by means thereof both the Motion of Light is propagated outwards, or from the solid Body to all imaginable Distance in Radiating Lines or Orbicular Pulses, with unimaginable Celerity : And also the Gravitation or Motion of Descent from all imaginable Distance towards the Radiating Body, of all solid Bodies, is caused and produced by the like Radiating Lines or Orbicular Pulses reversed, with an unimaginable Celerity and prodigious Power."

Conflicting with this, Hooke claimed :

"I could farther also prove, that every one of these distinct internal Motions of Bodies, as that of Light, and that of Sound, have distinct and differing Mediums, by which those Motions are communicated from the affecting to the affected Body : And so I conceive also that the Medium of Gravity may be distinct and differing both from that of Light, and from that of Sound."

The conception by Hooke of a universal gravitation and a law of inverse squares was the result of intuition. The statements were correct but Hooke did not have the mathematical talents of Newton to demonstrate their correctness. Moreover, Hooke did not seem to appreciate the need for such a mathematical demonstration, and this led him to charge Newton with plagiarism when the *Principia* finally appeared.

In 1684 Halley, who had discussed with Hooke and Wren a law of inverse squares, went to Newton in Cambridge to ask what planetary path would be described on the assumption of an inverse square law of distance. Newton's reply excited Halley, who pressed him to present the calculations. The result, the *Principia*, was published in 1687 at Halley's expense. Receiving the manuscript of this, Hooke had insisted that Newton insert an acknowledgement stating that the law of inverse squares was his own idea. Newton retorted furiously that Hooke was reducing the structure of the new universal mechanics, which was mathematical, to the level of dry calculation. In fact, Hooke did not appear to understand that Newton's work had gone beyond his own, for the conclusions were the same.

The industrious Hooke combined his scientific interests and duties with the position of City Surveyor. With Christopher Wren, a close friend, he assisted in the rebuilding of London, destroyed by the Great Fire. As well as Wren and Evelyn, Hooke drew up a plan for the City. This consisted of a system of straight streets crossing at right angles; but the plan was not accepted. The construction of canals and sewers, bridges and quays, was the responsibility of Hooke in the new civil architecture. The Royal College of Physicians, incorporating a magnificent octagonal theatre over its gate, was his own conception. This work began around 1671 and was completed within 7 years, but the building has not survived. Another of his structures, also lost during the nineteenth century, was the long Bedlam Hospital in Moorfields, started in 1674. It now seems that the Monument was the work, not of Wren, but of Hooke. The foundations were laid a year before Bedlam, and by May 1678 Hooke was at the top, experimenting on atmospheric pressure. Under the direction of Wren he assisted in the realization of the Royal Observatory at Greenwich, though it is not clear what credit is to be attributed to him. Wren often discussed St. Paul's with Hooke, who, according to one entry in his diary, made a suggestion concerning arches, which induced Wren to modify one of his models.

A well-known public figure, a frequent visitor to the coffee houses of London, Hooke has been vividly described by his early eighteenth-century biographer, Richard Waller:

"As to his Person he was but despicable, being very crooked, tho' I have heard from himself, and others that he was strait till about 16 Years of Age when he first grew awry, by frequent practicing, turning with a Turn-Lath, and the like incurvating

Exercises, being but of a thin weak habit of Body, which increas'd as he grew older, so as to be very remarkable at last: This made him but low of Stature, tho' by his Limbs he shou'd have been moderately tall. He was always very pale and lean, and laterly nothing but Skin and Bone, with a meagre Aspect, his eyes grey and full, with a sharp ingenious look whilst younger; his Nose but thin, of a moderate height and length; his Mouth meanly wide, and upper Lip thin; his Chin sharp, and Forehead large; his Head of a middle size. He wore his own Hair of a dark Brown colour very long and hanging neglected over his Face uncut and lank, which about three year before his Death he cut off, and wore a Periwig. He went stooping and very fast (till his weakness a few Years before his death hindred him) having but a light Body to carry, and a great deal of Spirits and Activity, especially in his Youth.

"He was of an active, restless, indefatigable Genius even almost to the last, and always slept little to his Death, seldom going to Sleep till two, three or four a Clock in the morning, and seldomer to Bed, often continuing his studies all Night, and taking a short Nap in the Day. His temper was Melancholy, Mistrustful and Jealous, which more increas'd upon him with his Years. He was in the beginning of his being made known to the Learned, very communicative of his Philosophical Discoveries and Inventions, till some Accidents made him to a Crime close and reserv'd. He laid the cause upon some Persons, challenging his Discoveries for their own, taking occasion from his Hints to perfect what he had not; which made him say he would suggest nothing till he had time to perfect it himself, which has been the reason that many things are lost, which he affirm'd he knew. He had a piercing Judgement into the Dispositions of others, and would sometimes give shrewd Guesses and smart Characters."

Robert Hooke, inventor of a universal joint and the wheel barometer, designer of instruments, was an excellent mechanic in a mechanical age. His versatility and skilful manipulation controlled the experimental research of the early Royal Society. Caught up in the enthusiasm and optimism of seventeenth-century science, his wild imagination soared to the most sublime of scientific truths and descended into crude and vague speculation. In his considerable body of writing there is something haphazard and disjointed, betraying a lack of insight and conviction, and for this, Hooke can never be regarded as a great mind.

FURTHER READING AND REFERENCE

Much of Hooke's work is readily accessible in:

GUNTHER, R. T., *Early Science in Oxford.* Vols. 6, 7, *The Life and Work of Robert Hooke*, 1930. Vol. 8, *The Cutler Lectures of Robert Hooke*, 1931. Vol. 10, *Tract on Capillary Attraction, 1661. Diary, 1688–93*, 1935. Vol. 13, *Micrographia*, 1938.

A paperback edition of *Micrographia* has been published by Dover, N.Y., 1961.

Another collection of Hooke's papers, not in Gunther, is:

WALLER, R., *The Posthumous Works of Robert Hooke,* 1705.

Hooke's criticism of Newton's paper on light and Newton's reply are contained in:

COHEN, I. B., and SCHOFIELD, R. E., *Isaac Newton's Papers and Letters on Natural Philosophy*, C.U.P., 1958.

For a recent biography, from which I have taken the details of Hooke's work as surveyor and architect, see:

ESPINASSE, M., *Robert Hooke*, London, 1956.

CHAPTER 6

ISAAC NEWTON, 1642–1727

D. W. Hutchings

One of the great difficulties in the study of the history of science is that of clearing one's mind of long-accepted and familiar ideas, so putting oneself on equal terms with early scientists. The ideas of "mass" and "force" are now so much part of us that it is only by a great stretch of the imagination that we can put ourselves in the place of scientists living in the first part of the seventeenth century. True that most of the concepts of modern mechanics were implicit in the work of Galileo, whose famous book on mechanics, *Dialogues Concerning Two New Sciences*, was published in Leyden in 1638, but his great contribution was the effect of his way of thinking rather than in the clarification of the concepts invoked. It could, however, be argued that by the time of Newton, Galileo's influence had been such that Newton himself could apply his intellect to problems *in* science rather than to establishing new attitudes and methods *for* science.

Let us then try to put ourselves back to the time of Newton's childhood. Although the Civil War was over and Charles I executed, England was not at peace with itself. It is probable that Newton's widowed mother living at Woolthorpe Manor in Lincolnshire went in fear of rough raiding parties after provision and plunder. And, when she married a second time and left young Isaac to live with his grandmother, it is perhaps not too difficult to imagine his excitement on being told that he would be sent away in order to go to school in Grantham.

It is much more difficult to put ourselves into the intellectual climate of that time. By the middle of the seventeenth century, as the former chapters tell, a great deal of work had been going on especially in astronomy, mathematics, and mechanics. But even to the intellectuals living then, it must have been extremely difficult to piece together the various ideas and make sense of them. There were many apparent contradictions and complications. Yet there was no shortage of exciting ideas, and at meetings of the Royal Society in England and the Académie des Sciences in France the natural philosophers of the day vigorously defended or criticized each other's work.

In these discussions an underlying idea was taking root—that the universe was a vast and intricate piece of machinery, or less naïvely, the universe could be likened to a wonderful clockwork. It was felt that a "model" of the universe could be put together much as a cathedral clock was put together, if only one knew which were the parts of the mechanism and the relationships between them. Of course, the seventeenth-century scientists were well aware of the difficulties of a rigorously mechanical explanation of the universe. By all appearances there were no physical connections between, say, the earth and the moon. But if the planets all move in a simple way, as Kepler and others had shown, it was reasoned that there must be a mechanism to make them move like this. Kepler himself believed in a simple *cause* producing the motion, and he sought that cause in mechanical terms. He thought of forces somehow radiating from the sun, like spokes from a wheel, pushing the planets around in their orbits. Descartes postulated that all space must be filled with some unobservable matter and that it was the swirling of this matter that carried the planets on their journeys. The important thing was to make accurate observations and measurements of the various parts of the mechanism, even if for the time being the exact nature of the mechanism was hidden.

Mechanics had begun as a study of problems connected with simple machines, like the wedge, pulley, and lever. Later more complicated machines with moving parts, like pumps and clocks, became the subject matter. The new question in the minds of men was whether the model of the universe worked according to the same rules as machines on earth.

An important task, before this question could be answered, was to try to clear up much of the vagueness about the rules of ordinary mechanics. As we have seen, both Galileo and Huygens made use of various mechanical concepts without actually defining them. What precisely was meant by "force", "mass", or "weight"? And what was meant by "impetus" or "impact"? But in spite of these difficulties, there was a growing realization that students of mechanics and astronomers were often dealing with the same kind of problems, and the application of mathematics to these problems was becoming the accepted practice. At meetings of the Royal Society in London and at the Académie des Sciences in Paris there were discussions about the effect of gravity on the motion of the planets.

Briefly the problem was this. Kepler's laws of planetary motion indicated that the planets moved in elliptical orbits according to definite mathematical principles. Galileo's analysis of the motion of a projectile "explained" curved motion in terms of straight-line motion. As we have seen, the concept of straight-line inertia was used by Huygens to work out his theory of centrifugal force. A body whirled in a circular orbit tends to continue in a straight line and can only be made to continue its circular motion if given an acceleration towards the centre. It is common experience that a stone whirled on the end of a string is only kept in its circular orbit by the pull of the string. Huygens solved the problem of the magnitude of the "centrifugal" force required to keep a body moving in a circle with constant speed.

Now, as planets move approximately in circular orbits, some force must act between them and the sun about which they rotate. Huygens was able to work out that for circular orbits the attractive force towards the sun would be an inverse-square one, following Kepler's third law. This was an important advance because it postulated a continuous force from the sun being needed to keep planets in circular paths. The exciting question of whether the same relationship would hold for elliptical orbits exercised a number of mathematicians and astronomers of the day. Hooke, Wren, and Halley failed to solve the problem, then Halley asked Newton to try it. Solving this problem, though spectacular, was only part of Newton's achievement.

NEWTON'S EARLY LIFE

Isaac Newton was born at Woolsthorpe in Lincolnshire on Christmas Day, 1642. Unlike the other seventeenth-century *savants* mentioned in this book, whose parents were rich or well connected, it is interesting to note that Newton came from a humble country family. His opportunity came when at the age of 12 he was sent to the King's School in Grantham. As the school was 7 miles from his home and too far from him to walk each day, he lodged in the town with an apothecary named Clark, who encouraged him to make things with his hands and to take an interest in chemistry. Newton showed no special promise at this time, and even after he entered Trinity College, Cambridge, in 1661, his immense powers lay dormant for three more years.

It was during his final undergraduate year that we first hear of him coming into contact with Isaac Barrow, who was the Lucasian Professor of Mathematics at Cambridge. Barrow evidently recognized Newton's promise and, as well as encouraging his mathematical studies, directed his attention to optics. Newton was soon reading works by Wallis, Descartes, and other advanced mathematicians of the day. He also became absorbed in experiments with lenses. He had always enjoyed making things, and regarded grinding his own lenses as a form of relaxation. During this period Newton began to grapple with refraction of light and mathematical work which eventually led to his binomial theory.

Yet Barrow had little time with his pupil, for in 1665, the year in which Newton took his degree, the university authorities decided that the students should be sent home because of the threat of the bubonic plague. That summer more than one-tenth of the population of London had died from this plague and there were signs of the disease spreading to parts of East Anglia.

Newton returned to the seclusion of Woolsthorpe. Without a college library and his optical apparatus, he was led to meditate on other problems. Whether or not the "apple incident" took place is by no means certain, though Newton's friend and biographer William Stukeley said that Newton had mentioned it to him. Also, Voltaire claimed that Newton was prompted to speculate about the cause of the fall of an apple while sitting in the orchard of his home. This led him further to ponder how far the apparent attraction of the earth would extend. Stones would continue to fall down the deepest mines or from the top of the highest cliffs or hills. Might not the same force of attraction extend as far as the moon and so explain how it *falls* towards the earth away from a straight path. Fortunately we have Newton's own account of how he first tackled this problem :

"And the same year I began to think of gravity extending to ye orb of the Moon, and having found out how to estimate the force with which a globe revolving within a sphere presses the surface of the sphere, from Kepler's Rule of the periodic times of the Planets being in a sesquialternate proportion of their distances from the centres of their Orbs I deduce that the forces which keep the Planets in their Orbs must be reciprocally as the squares of their distance from the centres about which they revolve : and thereby compared the force requisite to keep the Moon in her Orb with the force of gravity at the surface of the Earth, and found them answer pretty nearly."

The crucial calculation was like this. The distance of the moon's centre from the earth's centre is about sixty times the radius of the earth. If the inverse square law holds true, the attractive force of the earth on the moon must be $1/60 \times 1/60$ or $1/3600$ of what it is on the earth's surface.

An apple falls 16 feet in the first second after being dropped. So if the apple could be placed where the moon is, it would fall $16/3600$ feet in 1 second. But how does this compare with the true figure for the moon's fall in a second? It was possible to calculate the giant circle which the moon followed around the earth. Then, knowing that it takes $27\frac{1}{2}$ days for the moon to go round the circle once, it is easy enough to find how many feet per second the moon moves in this orbit.

Newton drew a scale diagram using the figures he had obtained and found exactly how much the moon falls off from the horizontal in one second; that is, how many feet it drops towards the earth in a given time. As we have seen, he found that the answer tallied "pretty nearly", and later, when more accurate measurements for the size of the earth were available, the result was almost perfect.

In view of this, there has been much speculation as to why Newton did not publish the result of his calculations at once. It may have been because he had been obliged to make a number of simplifying assumptions. For example, he treated the earth, the moon, and the apple as though the whole of each body was concentrated at what we should now call their centres of gravity. Also, in calculating the force between the earth and the apple, the size of the earth is gigantic compared with the apple or the distance between the two bodies, and this raises a number of difficulties. But, whatever the reasons, about another 20 years were to pass before the law of gravitation was announced.

It was during this period at Woolsthorpe also that Newton evolved his "method of fluxions" or the "calculus". He needed a way of dealing with varying velocities. Constant velocity of, say, a motor-car can be expressed as distance/time. If s is the distance travelled in time, t, the velocity will be s/t. The velocity will always be the same however great or small s or t may be. But if the velocity is not constant but changing, its value at any instant can only be found by taking a time so short that it can be assumed that the velocity will not have changed appreciably. Newton argued that when s and t were infinitesimally small, the quotient s/t gave the velocity at that instant. He wrote this quantity as s, a notation soon to be superseded by that of Leibnitz. Of course, any two quantities which depend on each other can be treated in the same way : the rate of variation of x and y is written as dx/dy in modern notation.

The reverse process now known as "integration", enables the quantity itself to be estimated from a knowledge of its rates of change. This was needed by Newton in order to tackle such problems as the calculation of the attraction of a whole sphere from the attractions of each of its myriads of particles.

On returning to Cambridge in 1677 Newton was elected to a minor fellowship at Trinity, and in the following year he became a senior Fellow of the College. When Barrow retired in 1669, Newton, now aged 27, was elected as the new Lucasian Professor of Mathematics. The official duties were not onerous—only one lecture per week for one of the university terms. This must have suited Newton's temperament and he was able to devote himself to his studies and his experiments. His main interest at this time was optics and, as so often happens in the history of science, he was first impelled to study the nature of light in order to solve a technical problem. In his attempts to make a more efficient telescope, he studied the properties of rays of light when passed through lenses and prisms. An interesting account of some of these early experiments was described in a letter to the newly formed Royal Society (printed in the *Philosophical Transactions* of February 1672):

"Having darkened my chamber and made a hole in my window-shuts to let in a convenient quantity of the Sun's light, I placed my Prisme at his entrance, that it might be thereby refracted to the opposite wall. It was at first a very pleasing divertisement to view the vivid and intense colours produced thereby; but after a while applying myself to consider them more circumspectly, I became surprised to see them in an *oblong* form, which, according to the received laws of Refraction, I expected should have been *circular*.

"They were terminated at the sides with straight lines, but at the ends, the decay of light was so gradual that it was difficult to determine justly what was their figure; yet they seemed semi-circular.

"Comparing the length of this coloured *Spectrum* with its

breadth, I found it about five times greater; a disproportion so extravagant that it excited me to a more than ordinary curiosity of examining from whence it might proceed. I could scarce think that the various *Thickness* of the glass, or the termination with shadow or darkness could have any influence on light to produce such an effect; yet I thought it not amiss, first to examine those circumstances, and so tryed what would happen by transmitting light through parts of the glass of divers thicknesses, or through holes in the window of various bignesses, or by setting the Prisme without so that the light might pass through it and be refracted before it was terminated by the hole. But I found none of those circumstances material. The fashion of the colours was in all these cases the same."

One of the drawbacks of the early refracting telescopes arose from the fact that "white" light is made up of all the colours of the rainbow, and when a beam of light is passed through a glass lens the various colours are brought to focus in different places. Stars viewed through an ordinary refractor appeared to be surrounded by false colour. Once Newton had observed the effect of a prism on a shaft of light, he recognized that this imperfection resulted from the nature of light itself.

His method of minimizing the blurring of objects seen through a telescope was ingenious : he invented a "reflecting" telescope containing a concave mirror instead of the usual lens. This mirror formed an image inside the tube, where it is reflected at right-angle by a small plane mirror or "flat" so that it could be viewed by an eyepiece let into the side of the tube (Fig. 6.1).

The crucial experiment by which Newton satisfied himself that sunlight is not pure or homogeneous but consists of rays of differing "refrangibilities" was described in his letter to the Royal Society :

"I took two boards, and placed one of them close behind the Prisme at the window, so that the light might pass through a small hole, made in it for the purpose, and fall on the other board, which I placed at about 12 feet distance, having first made a small hole in it also, for some of that incident light to pass through. Then I placed another Prisme behind this second board, so that the light trajected through both boards might pass through that also, and be again refracted before it arrived at the wall. This done, I took

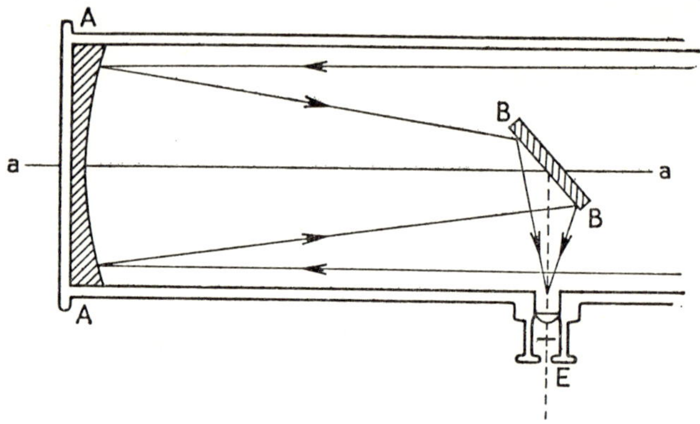

Fig. 6.1. Newton's reflecting telescope.

the first Prisme in my hand and turned it to and fro slowly about its *Axis*, so much as to make the several parts of the Image, cast on the second board, successively pass through the hole in it, that I might observe to what places on the wall the second Prisme would refract them."

By this method Newton isolated small regions of the spectrum and subjected them to further refraction. A beam of each colour in turn was passed through the hole in the first board, sharply delineated by the hole in the second board, and refracted by the second prism on to the wall beyond.

It would be wrong to give the impression that Newton was the first to suspect that there was a connection of some kind between the nature of light and the nature of colour. Antonio de Dominis, Archbishop of Spalatro, had suggested in 1611 that the colours of the rainbow were due to light reflected from raindrops having to pass through different thicknesses of water. This was improved upon by Descartes, who related the colour with the refrangibility. Nor was Newton the first to pass white light through a prism.

Newton not only extended previous experiments using lenses and prisms but sought the causes of colours of thin plates, as in bubbles and in other films. In one experiment he pressed a glass prism on to a lens of known curvature and found that the colours were formed in circles, since called "Newton's rings". By carefully measuring the rings and comparing the measurements with the estimated thicknesses of the air films in various positions, Newton supposed that light of each definite colour was subject to "fits" of easy transmission and easy reflection. This experiment was repeated using light of one colour only, when alternate light and dark rings were produced. He also carried out experiments with very narrow beams of light which are bent at the sharp edges of obstacles so that the shadows are larger than one might expect if light travels strictly in straight lines; moreover, fringes of colour are formed. In an ingeniously conceived experiment, Newton showed that the bending is increased if light is passed through a very narrow slit formed between two knife edges. He was able to measure the breadths of such slits and the varying angles of deflection.

Another effect, actually first observed by Huygens, gave a further clue to the nature of light. When light is passed through certain minerals like Iceland Spar the refracted rays exhibit an unusual property, now called "polarization". If one of these refracted rays is directed into another spar crystal, it is found that it will pass through when the axes of both crystals are parallel but will not pass when the axis of the second crystal is at right-angles to that of the first. Newton saw that this meant that a ray of light was somehow unsymmetrical, being different on different sides. These results are important from a historical point of view because they show that Newton realized that light must have wave-like properties as well as particle properties. He commented on these in his letter to the Royal Society:

"Nothing more is requisite for putting the Rays of Light into Fits of easy Reflexion and easy Transmission, than that they be small Bodies which by their attractive Powers, or some other Force, stir up Vibrations in what they act upon, which Vibrations being swifter than the Rays, overtake them successively, and agitate them so as by turns to increase and decrease their Velocities, and thereby put them into those Fits. And lastly, the unusual Refraction of Island Crystal looks very much as if it were perform'd by some kind of attractive virtue lodged in certain sides both of the Rays, and of the Particles of the Crystal."

People who are in the public eye invariably come in for criticism. Newton was no exception and he was bewildered and pained by the controversies following the publication of his letters in the *Philosophical Transactions of the Royal Society*. Several of the most distinguished scientists of the day regarded Newton's work as an attack on their own theories concerning the nature of light and colour. In particular, Hooke and Huygens had independently assumed the existence of *subtle matter* to transmit vibrations in the working out of their respective "wave theories". To Newton the fact that light travels in straight lines, forming sharp shadows, seemed to demand a corpuscular theory.

After answering their criticisms one after the other, he reached a point when he refused to be drawn into any more conflict. He felt that much of the argument was pointless as the experimental facts spoke for themselves. He fought shy of publicity and retired into a cloistered academic life. Thus it was that Newton, then in his early thirties, was able to devote himself to a study of theology and chemistry. His assistant Humphrey Newton, who may have been a relative, recorded that :

"He very rarely went to bed before two or three of the clock, sometimes not till five or six . . . especially at spring and fall of leaf, at which time he used to employ about six weeks in his elaboratory, the fire scarce going out night or day, he sitting up one night and I the other, till he had finished his chemical experiments."

Although Newton wrote no book on chemistry, it is possible that he spent more time on chemical experiments than the other studies which made him famous. He seems to have been particularly interested in metals and alloys and his writings showed an insight into the structure of matter and the affinity of one chemical substance for another beyond that of other chemists of his day.

PRINCIPIA

Newton paid his first visit to the Royal Society in 1675 and at subsequent meetings he again came into conflict with Hooke. This put him on his mettle for a number of the fellows were keenly interested in gravity and its effect on planetary motion. At any rate, Newton was to take up again some of those problems which he had laid aside.

Even so, there was little to indicate that he was contemplating his monumental work *Principia* and the circumstances which led to the publication of this remarkable book are interesting.

Hooke, Wren, and the astronomer Halley had been discussing whether a planet moving under attraction in accordance with the inverse square relation, as suggested by Kepler's third law, would describe an ellipse in accordance with his first law. But they had been unable to give a mathematical proof of the relationship. Wren offered a prize "in the form of a book of the value of forty shillings" to whoever would bring him a convincing solution of the problem within a couple of months. As the months rolled by and no proof was forthcoming, Halley became impatient and he decided to go to Cambridge and ask Newton if he could do it. To his surprise and delight, Newton told him that he had already solved the problem. Halley must have been very disappointed when Newton failed to find the proof among his papers but, a few months later, when Newton provided two solutions, Halley realized that there was a scientific achievement of the greatest importance.

Soon after Newton had sent these proofs of the elliptic orbits to Halley, he began to set down his ideas on mechanics in a little book *De Motu corporum* (*On the Movements of Bodies*). This he also sent to Halley, who at once resolved to ask Newton to write a detailed treatise embodying his discoveries. Newton set to work in 1685 and, after 17 months of incredible labour, the *Principia*, which has been described as the most extraordinary production of the human mind, was ready for the printer.

The Minutes of the Royal Society for 28 April 1686 recorded :

"Dr. Vincent presented to the Society a manuscript treatise entitled *Philosophiae Naturalis Principia Mathematica*, and dedicated to the Society by Mr. Isaac Newton, wherein he gives a mathematical demonstration of the Copernican hypothesis as proposed by Kepler, and makes out all the phenomena of the celestial motions by the only supposition of a gravitation towards the centre of the sun decreasing as the squares of the distance therefrom reciprocally."

This actually refers to the first of three "books" which comprise *Principia.* The whole work of some 250,000 words was written in Latin and is, of course, highly mathematical and difficult to read. A reliable English translation was made by Andrew Motte in 1729 and is the source of the quotations here. In the preface, Newton gave a clear outline :

"Since the ancients (as we are told by Pappus) esteemed the science of mechanics of the greatest importance in the investigation of natural things, and the moderns, rejecting substantial forms and occult qualities, have endeavoured to subject the phenomena of nature to the laws of mathematics. I have in this treatise cultivated mathematics as far as it related to philosophy . . . for the whole burden of philosophy seems to me to consist of this— from the phenomena of motions to investigate the force of nature, and then from these forces to demonstrate the other phenomena, and to this end the general propositions of the first and second Books are directed. In the third Book I give an example of this in the explication of the System of the World; for by the propositions mathematically demonstrated in the former Books in the third I derive the celestial phenomena the forces of gravity with which bodies tend to the sun and several planets. Then from the forces, by other propositions which are also mathematical, I deduce the motions of the planets, the comets, the moon, and the sea. . . ."

Book I is concerned with mechanics and begins with a set of definitions—mass, momentum, inertia, centripetal force. The movements of bodies moving in any curve—circle, ellipse, parabola, or hyperbola—about a stationary body situated at a focus are considered as taking place in an empty space, with no resistance to their motion. By assuming that the particles of the bodies are massive but of negligible volume, Newton shows that in each case an inverse square law of attractive force will account for the motions. Here too are the famous three Laws of Motion :

Law I : Every body preserves in its state of rest or of uniform motion in a straight line, except in so far as it is compelled to change that state by impressed forces.

Law II : Change of motion (i.e. rate of change of momentum) is proportional to the moving force impressed, and takes place in the direction of the straight line in which such force is impressed.

Law III : Reaction is always equal and opposite to action; that is to say, the actions of two bodies upon each other are always equal and directly opposite.

In the second book Newton considers movements in various resisting media, as bodies moving in water or air. Among the many cases dealt with is the motion of pendulum in air—a subject of great interest at that time and for which there was a good deal of well-substantiated experimental evidence.

This brings out an interesting point : the philosophic method adopted by Newton in *Principia* is essentially deductive. He began by stating his assumptions much in the same way as Euclid began with a number of "axioms". Then by series of rigorous mathematical steps new deductions were made. His four rules of reasoning are stated in the third of the books :

"I. We are to admit no more causes of natural things than such as are both true and sufficient to explain their appearances. To this purpose the philosophers say that Nature does nothing in vain, and more is vain when less will serve; for Nature is pleased with simplicity, and affects not the pomp of superflous causes.

"II. Therefore to the same natural effects we must, as far as possible assign the same causes. As to respiration in a man and in a beast; the descent of stones in Europe and in America; the light of our culinary fire and of the sun; the reflection of light in the earth, and in the planets.

"III. The qualities of bodies, which admit neither intention nor remission of degrees, and which are found to belong to all bodies whatsoever. For since the qualities of bodies are only known to us by experiments, we are to hold for universal all such as universally agree with experiments . . . we are certainly not to relinquish the evidence of experiments for the sake of dreams and vain fictions of our own devising. . . .

"IV. In experimental philosophy we are to look upon propositions collected by general inductions from phenomena as accurate or very nearly true, notwithstanding any contrary hypotheses that may be imagined, till such time as other phenomena occur, by which they may either be made more accurate, or liable to exceptions."

The first of these rules has been called a "Principle of parsimony" since it states that Nature is essentially simple and that we should not introduce more hypotheses than are sufficient and necessary for the explanation of the observed facts. The second and third rules are "principles of unity" : as far as possible, similar effects must be assigned to the same cause and properties common to all those bodies within reach of our experiments are to be assumed, even if only tentatively, to all bodies in general. Thus we assume that the laws of mechanics which apply to falling stones or cannon balls would apply also to the moon or the planets.

The fourth rule states the scientist's faith that hypotheses based on a range of experimental findings can be regarded as exactly or approximately true until further phenomena or experiments show that they may be corrected or are liable to exceptions.

There was no shortage of ingenious hypotheses to account for the movements of the planets or the nature of light. Descartes had suggested that the planets were swept around in their orbits by the swirling of a medium of minute particles which he had postulated as filling all space. Newton, in effect, examined the mathematical consequences of these ideas. He pointed out that to drag a body, a fluid must offer resistance to its motions and then demonstrated that the laws of motion of a body swept around by a vortex of fluid would not account for Kepler's laws :

"If a solid sphere, in a uniform and infinite fluid, revolves about an axis given in position with a uniform motion, and the fluid be forced round by only this impulse of the sphere; and every part of the fluid perseveres uniformly in its motion : I say, that the periodic times of the parts of the fluid are as the squares of the distances from the centre. . . .

"Hence it is manifest that the planets are not carried round in corporeal vortices; for according to the *Copernican* hypothesis, the planets going round the sun revolve in ellipses, having the sun in their common focus; and by radii drawn to the sun describe areas proportional to the times. But the parts of a vortex can never revolve with such a motion."

Other calculations showed that Descartes's hypothesis was untenable on several other counts. Unless there is some source of the motion, the swirling medium would gradually slow down and eventually come to rest. This in turn would mean that the length of the year would vary. Also, neighbouring bodies, each rotating, would tend to move each other.

Book II contains many other interesting topics. In considering the frictional forces operating when bodies move in a medium like air or water, Newton applied himself to the problem of the shape that would meet the least resistance. He mentions that the result "may be of use in the building of ships". This calculation involved what we should now call the calculus of variations. Another important piece of mathematics introducing new fundamental ideas concerned with wave motion is also to be found in the second book.

OF THE SYSTEM OF THE WORLD

However, it is in the third book of *Principia* that Newton does no less than to "demonstrate the frame of the System of the World". He announces his discoveries concerning gravitation and gives a mathematical explanation of the way in which the Earth and the other planets move. After summarizing his foregoing results, he works out in detail the motions of the satellites of Jupiter, Saturn, and the Earth, and of the planets round the sun. The underlying *idea* of universal gravitation, that every particle attracts every other particle with a force diminishing with the square of the distance, is applied in each case. In demonstrating that the law of gravitation holds for heavenly bodies as well as moving objects on the Earth, Newton provides a description of *how* things happen and not why they happen. This description is so elegant and, as Newton went on to show, capable of such wide application, that for more than two hundred years it remained unchallenged.

Newton was able to calculate the masses of the sun and of the planets from the mass of the earth; to account for the flattening of the earth at the poles, the variation of the gravitational force over the surface of the earth, and the disturbance of the moon's orbit owing to the sun. Another remarkable calculation accounted for the tilt of the earth and the so-called precession of the equinoxes. Newton shows that they are members of the solar system obeying the same laws. The famous comet of 1682, whose path had been carefully observed by Halley, indicated to him a period of reappearance of about 75 years if the comet, although an eccentric member of the solar system, were to obey the laws of mechanics. Its return in 1756 and twice since, is further evidence of the power of Newton's reasoning (Fig. 6.2).

Another important phenomenon explained by Newton through the application of the law of gravitation was that of the tides. He reasoned that the moon (and to a lesser extent the other heavenly

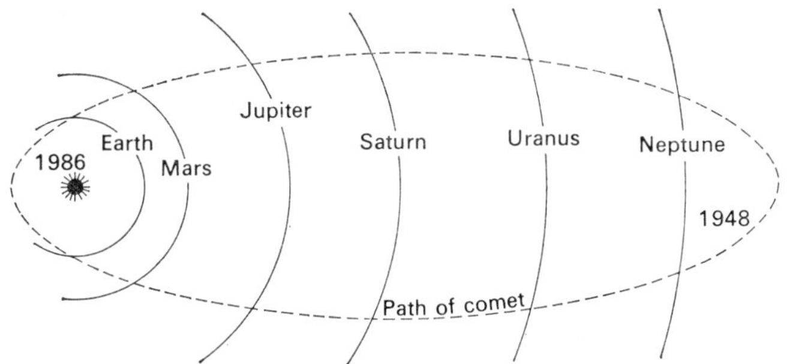

FIG. 6.2. Path of Halley's comet.

bodies) would pull on the nearest part of the ocean, and so tend to "heap up the waters". Of course, there are many complicating factors but the essence of Newton's calculation is as follows :

"Since the waters attracted by the sun's force rise to the height of 1 foot and $11\frac{1}{30}$ inches, the moon's force will raise the same to the height of 8 feet $7\frac{5}{22}$ inches; and the joint forces of both will raise the same to the height of $10\frac{1}{2}$ feet; and when the moon is in its perigee to the height of $12\frac{1}{2}$ feet, and more, especially when the wind sets the same way as the tide. And a force of that amount is abundantly sufficient to produce all the motions of the sea, and agrees well with the ratio of these motions. . . ."

In all of these applications, then, Newton's theory involved a synthesis of physical phenomena which had formerly been treated separately; for example those happenings studied by astronomers, and those concerned with motion in everyday experience on earth. And the very form or style in which the *Principia* was written underlined this main idea. As we have said, it resembled Euclid's approach to geometry with definitions, axioms and theorems. However, there is an important difference. Whereas Euclid's axioms were supposedly self-evident principles about relationships between abstract mathematical concepts, Newton's axioms consisted of observable and respectable facts found by experiment. Put another way, Newton's synthesis by employing empirical data and abstract mathematical relations led to new verifiable observations. The success of his method seemed to suggest that the universe is orderly and determined—that all events take place in a definite order.

Newton himself believed that the universe was more than a very complicated machine but that God ordered the movements of its parts :

"This most beautiful system of the sun, planets, and comets, could only proceed from the counsel and dominion of an intelligent and powerful Being. And if the fixed stars are the centres of other like systems, these, being formed by the like wise counsel, must be all subject to the dominion of One; especially since the light of the fixed stars is of the same nature with the light of the sun, and from every system light passes into all the other systems; and lest the systems of the fixed stars should, by their gravity, fall on each other mutually, he hath placed those systems at immense distances one from another. . . . This Being governs all things, not as the soul of the world, but as Lord over all. . . . And thus much concerning God; to discourse of whom from the appearances of things does certainly belong to Natural Philosophy."

LIFE IN LONDON

Early in 1689 Newton was elected Member of Parliament by the University of Cambridge. Later that year, after the flight of James II, Parliament was summoned by the Protestant William and Mary. Macaulay, in a vivid description of the event, commented on Newton's presence : "Among the crowd of silent members appeared the majestic forehead and pensive face of Isaac Newton . . . he sat there, in his modest greatness, the unobtrusive but unflinching friend of civil and religious freedom."

His new duties of course brought him to London quite frequently and he made many new acquaintances and friends, like John Locke, the philosopher, and Samuel Pepys, the famous diarist. But, in spite of his fame, all was not well with Newton about this time. He went through an unhappy period, often sleepless at night, and sometimes convinced that even his most intimate friends were against him—one might say that he was on the verge of a nervous breakdown. However, in March 1696, he was offered the post of Warden of the Mint and so came to live in London and began a period of public life rather different from the secluded life of a Cambridge college. He was a great success at the Mint and, in spite of many difficulties and a certain amount of opposition, he succeeded in a complete reorganization of the system of coinage. Again quoting Macaulay: "The ability, the industry and strict uprightness of the great philosopher, speedily produced a complete revolution throughout the department which was under his direction."

And, when in 1699, he was appointed to the superior position of Master of the Mint, he began to enjoy a period of greater peace and contentment. He had a house in Jermyn Street, his niece Catherine Barton acted as housekeeper, and he found great comfort and pleasure from the friendship and esteem of visitors from all over the world. In 1703 Newton was elected President of the Royal Society, a position he fulfilled with enthusiasm until his death almost a quarter of a century later.

OPTICKS

It was in 1704, the year after his former adversary Hooke died, that Newton published his great work on light. Written in the English language, it was called *Opticks: or a Treatise on the Reflexions, Refractions, Inflexions and Colours of Light*. A translation into Latin was made 2 years later. It may well have been the thought of more bitterness with Hooke that deterred Newton from publishing his work on light. After all he was in his sixtieth year and most of his speculation and experimentation on optics dated back to his time as a young Fellow at Trinity. There is no doubt that he wished to avoid any unpleasantness and in the preface of *Opticks* he says as much: "To avoid being engaged in Disputes about these Matters, I have hitherto delayed the printing, and should still have delayed it, had not the Importunity of Friends prevailed upon me."

Actually, the first part of the book deals with much that he had already published: experiments on the prism, coloured images produced by lenses, and the reflecting telescope. He then proceeds to discuss the nature of colour. He demonstrated by many different experiments that the colours "of" ordinary objects depend upon the kind of light falling on them. He says:

"The Colours of all natural Bodies have no other origin than this, that they are variously qualified to reflect one sort of light in greater plenty than another. And this I have experimented in a dark room by illuminating those bodies with uncompounded light of divers colours. For by that means any body may be made to appear any colour. They have there no appropriate colour, but ever appear of the colour of the light cast upon them, but yet with this difference, that they are most brisk and vivid in the light of their own daylight colour."

By "uncompounded light" is meant the light from one part of the spectrum only. And in this way Newton made clear what every schoolboy now knows, that a red object looks red because it absorbs all of the colours of the spectrum but lets the red out again. In other words, Newton was the first to explain how the colours of material bodies are related to the spectrum and are a consequence of the compound nature of light. Also, how the subtraction and addition of colours, as in the mixing of paints or the illumination of coloured objects with light of a particular colour, made sense.

He also includes a clear explanation of the rainbow. As we have seen, Descartes and others had given a general explanation of the phenomenon. By his experiments with prisms and by reconstituting white light by bringing together the coloured rays, Newton was able to clear up the subject. That the colours are due to the refraction of light rays entering minute raindrops and, after reflection inside the drop, are refracted out again. As in the case of the prism, the refraction by the water droplets is accompanied by a separation of the colours.

Also in the *Opticks*, Newton gives extensive consideration to two phenomena which proved to be of fundamental importance in the study of light: the colours of thin plates, such as soap bubbles or thin films of air between two plates of glass pressed together; and the fact that light bends slightly, but distinctly, around the edges of opaque objects illuminated by a very small source of light.

He was led to his investigation of the first phenomenon by an observation which might easily have been overlooked. He noticed a black spot at the point where two slightly convex prisms touch when pressed together. By slightly altering the position of the prisms, he found that circles of colour appeared. Although previously observed by Boyle and Hooke, these circles are now known as "Newton's rings". One experiment is described as follows:

"I took two Object glasses, the one a Plano-convex for a fourteen foot Telescope, and the other a large double Convex for one of about fifty foot; and upon this, laying the other with its plane side down wards, I pressed them slowly together, to make the colours successively emerge in the middle of the circles, and then slowly lifted the upper glass from the lower to make them successively vanish again in the same place."

Newton went far beyond his predecessors in making a very careful "quantitative" study of the ring phenomenon, calculating accurate values for the film thicknesses. He observed that a given colour of light was alternately reflected and transmitted with increasing film thicknesses, also that each colour was found to have a different thickness of film at which the first bright reflection was observed.

These facts lead to the inescapable conclusion that there is something periodic in the nature of light. Where Newton differed from Hooke and Huygens is that he attributed the periodic disturbances not to the nature of light itself but to "fits" of alternate reflection and transmission brought on by the medium, such as a thin film of air, through which it was passing :

"The returns of the disposition of any ray to be reflected, I will call its *Fits of easy Reflection*, and those of its disposition to be transmitted its *Fits of easy Transmission*, and the space it passes between every return and the next return, the *Interval of its Fits*.

"What kind of action or disposition this is; Whether it consists in a circulating or a vibrating motion of the Ray, or of the Medium, or something else, I do not here enquire. Those that are average from assenting to any new Discoveries, but such as they can explain by an Hypothesis, may for the present suppose, that as Stones by falling on Water put the Water into an undulating Motion, and all Bodies by percussion excite vibrations in the Air; so the Rays of Light, by impinging on any refracting or reflecting Surface, excite vibrations in the refracting or reflecting Medium or Substance, and by exciting them agitate the solid parts of . . . the Body . . . and cause the Body to grow warm and hot; that the vibrations thus excited are propagated in the reflecting or refracting medium or Substance, much after the manner that vibrations are propagated in the Air for causing Sound, and move faster than the Rays so as to overtake them; and that when any ray is in that part of the vibration which conspires with its motion, it easily breaks through a refracting Surface, but when it is in the contrary part of the vibration which impedes its Motion, it is easily reflected. . . . But whether this Hypothesis be true or false I do not here consider. I content myself with the bare Discovery that the Rays of Light are by some cause or other alternately disposed to be reflected or refracted. . . ."

Thus Newton warns about too easily jumping to conclusions. Basing himself on such experimental evidence as he had at his disposal, he shows that it was quite easy to invent a hypothesis which would be plausible enough : the periodicity could be accounted for by supposing a secondary effect in the medium rather than making periodicity an intrinsic property of the light rays. But as to the validity of such a hypothesis, he was not prepared to say.

This brings out another interesting point. In contrast to the method in *Principia*, which is almost entirely "deductive", Newton's work on light is a splendid example of "inductive" method. Most of *Opticks* is descriptive and experimental, always guarding against "hypotheses" that attempt to "explain" the observed phenomena.

Newton was insisting that scientific investigation should always be carried out in a definite order, always beginning with observation and experiment. The empirical data so collected, when generalized, are treated as the effects of certain universal causes. Then, and only then, can the order be reversed, using causes as principles from which new relationships can be derived. And, as we have seen, mechanics and astronomy had reached an advanced state so that Newton could state his principles and generalizations and then make deductions from them. But, studies of the nature of light had not yet, in Newton's view, reached this point. At the end of *Opticks* he says:

"When I made the foregoing Observations, I designed to repeat most of them with more care and exactness, and to make some new ones for determining how the Rays of Light are bent in their passage by Bodies. . . . But I was then interrupted, and cannot think of taking these things into farther Consideration. And since I have not finished this part of my Design, I shall conclude with proposing only some Queries, in order to farther the search to be made by others."

Then in the very considerable appendix to *Opticks* Newton permitted himself to speculate. These 31 *Queries* are far more than a list of interesting questions or suggestions for further lines of work. Coming, as they do, towards the end of Newton's scientific life, here we have the benefit of his powerful intellect contemplating some of the most puzzling questions about the form of the universe. Ranging over the nature of light, heat, electricity, magnetism and the structure of matter, many of the *Queries* provide very remarkable links between seventeenth century views and modern concepts. Perhaps the most interesting of these is the last *Query*, number 31, which is some 38 pages long. It includes the definite suggestion that the forces which hold atoms together are electrical in nature. When it is recalled that the static electricity produced by friction was the only kind of electricity known at that time, Newton's conjecture is the more remarkable. Referring to "the small particles of bodies", he speculates, "perhaps electrical Attraction may reach to such small distances, even without being excited by Friction".

LAST YEARS

In 1705, the year after the first edition of *Opticks*, Queen Anne, the Prince Consort, George of Denmark, and many members of the Court visited Cambridge. On this splendid occasion the Queen conferred a knighthood on Newton. He was the first scientist to be so honoured and clearly demonstrates the esteem with which he was now held.

The latter 20 years of Newton's life—he was 85 when he died—were very active. He remained Warden of the Mint and continued to discharge this important duty with great efficiency. As President of the Royal Society he presided over its affairs at nearly every one of its weekly meetings. He was also concerned with publication or new editions of his writings. There was a great demand for *Principia*—there were three editions in England during his lifetime as well as two more which were published in Amsterdam. As well as reprints in the original Latin, there were many translations into other languages. The first English translation by Andrew Motte in 1729, was made after Newton's death.

Among the smaller works published during his lifetime are two mathematical treatises *De Analysi*, of which Newton had given a handwritten copy to Isaac Barrow after his stay at Woolsthorpe, and *Arithmetica universalis*, concerned mainly with algebra. His lectures on optics were published in book form, *Optical Lectures*. There was also his book on religious subjects.

Recent research has given greater insight into Newton's relationships with his contemporaries. He was undoubtedly very touchy and testy, and there are many examples of his uncharitable behaviour to those who seemed to oppose his views. On the other hand, he gave great encouragement to others, particularly younger men. Although he became very rich, he continued to live relatively simply. He was generous with his money, and often careless with it.

Although he was continually acclaimed and honoured, even revered, Newton was modest about his own achievements. He said of himself: "I do not know what I may appear to the world; but to myself I seem to have been only a boy, playing on the seashore, and diverting myself in now and then finding a smoother pebble or a prettier shell than ordinary, while the great ocean of truth lay all undiscovered before me."

FURTHER READING AND REFERENCE

(a) General Histories

Butterfield, H., *Origins of Modern Science*, London, 1949.

Hall, A. R., *The Scientific Revolution, 1500–1800*, London, 1954.

Wolf, A., and McKie, D., *A History of Science, Technology and Philosophy in the Sixteenth and Seventeenth Centuries*, London.
This is a large and detailed work, recently revised. It is well illustrated.

(b) Special Aspects

Andrade, E. N. da C., *Sir Isaac Newton*, London, 1954. A very useful condensation of Newton's life and works, now issued in paperback.

Bell, A. E., *Newtonian Science*, London, 1961. A readable book in which the intellectual achievements of the scientific revolution are considered in the light of humanistic values.

Magie, W. F., *A Source Book of Physics*, London and New York, 1935. Translated excerpts from original works, with biographical and explanatory notes.

Toulmin, S., and Goodfield, J., *The Fabric of the Heavens*, London, 1960. This first book of a series of 4 volumes on the "Ancestry of Science" examines the development of astronomy and dynamics, and the contribution these sciences have made to cosmological ideas.